Social Movements Contesting Natural Resource Development

W0113281

Presenting a broad range of case studies, this book explores rural social movements contesting natural resource development initiatives.

Natural resource development takes multiple forms, including infra-structure corridors, mines, dams, resource processing plants and pipelines. Many of which are driven by economic valuations, whilst social and environmental effects are given limited consideration. In this volume the authors discuss the emergence, process and outcomes of social movements with respect to these natural resource development projects, including examples of confrontation seeking to either block developments or promote alternative development approaches, such as agritourism. The examples taken from Africa, Asia, North America, Europe and Latin America demonstrate the diversity of struggles stimulated by natural resource development, including both immediate and longer-term effects, repertoires of action, political and cultural work. Taken together the case studies provide a rich overview of current movements engaged in resisting the neoliberal agenda of global resource exploitation.

This book will be key reading for scholars interested in social movements, natural resource development, environmental policy and development studies. It will also be of interest to activists engaged in mobilizations stimulated by natural resource development projects.

John F. Devlin is Associate Professor, School of Environmental Design and Rural Development, University of Guelph, Ontario, Canada.

Earthscan Studies in Natural Resource Management

For more information on books in the Earthscan Studies in Natural
Resource Management series, please visit the series page on the Routledge
website: www.routledge.com/books/series/ECNRM/

Social Movements Contesting Natural Resource Development

Edited by John F. Devlin

Routledge
Taylor & Francis Group

LONDON AND NEW YORK

earthscan
from Routledge

First published 2020
by Routledge
2 Park Square, Milton Park, Abingdon, Oxon OX14 4RN

and by Routledge
605 Third Avenue, New York, NY 10017

First issued in paperback 2021

Routledge is an imprint of the Taylor & Francis Group, an informa business

British Library Cataloguing-in-Publication Data
A catalogue record for this book is available from the British Library

Library of Congress Cataloging-in-Publication Data
Names: Devlin, John F, editor.
Title: Social movements contesting natural resource development / edited by John F Devlin.
Description: Abingdon, Oxon ; New York, NY : Routledge, 2020. | Series: Earthscan studies in natural resource management | Includes bibliographical references and index.
Identifiers: LCCN 2019029748 (print) | LCCN 2019029749 (ebook) | ISBN 9781138064737 (hardback) | ISBN 9781315160139 (ebook)
Subjects: LCSH: Environmental protection–Citizen participation–Case studies. | Mineral industries–Social aspects-Case studies. | Conservation of natural resources–Social aspects–Case studies. | Natural resources management–Social aspects–Case studies. | Community-based conservation–Case studies.
Classification: LCC GE170 .S669 2020 (print) | LCC GE170 (ebook) | DDC 333.71/5–dc23
LC record available at https://lccn.loc.gov/2019029748
LC ebook record available at https://lccn.loc.gov/2019029749

ISBN 13: 978-1-03-208746-7 (pbk)
ISBN 13: 978-1-138-06473-7 (hbk)

Typeset in Sabon
by Wearset Ltd, Boldon, Tyne and Wear

I dedicate this book to Dr. Nonita Tumulak Yap, my wife and colleague. During 40 years of academic and social movement work she introduced me to international development and environmental issues from Australia to Zimbabwe and many countries in between. Nonita suffered a stroke and passed away while the book was being edited. She was my earth and my sky ... we struggle on ... *a luta continua.*

Contents

Figures

Contributors

Brock Bersaglio is a Postdoctoral Fellow in the Department of Geography at the University of Sheffield and an Affiliated Researcher with the East African Institute at Aga Khan University. Focusing on East and Southern Africa, his work investigates the implications of biodiversity conservation for land rights, livelihoods, and justice. His work also engages with natural resource governance and the politics of rural development more broadly.

Ismo Björn is a Senior Researcher at Karelian institute, University of Eastern Finland and Adjunct Professor in Finnish and Environmental History at UEF and University of Turku. He has studied environmental history, especially forest and mining history, sport history (today ski jumping and Finnish baseball) and the politics of memory on the Finnish-Russian border. He was born in Outokumpu beside the Outokumpu mine and its waste area. He has written several books about local Finnish history.

Angie Carter is an Assistant Professor of environmental and energy justice at Michigan Technological University. She studies agrifood systems sustainability, environmental problems, and social change through community-based research.

John F. Devlin is an Associate Professor in the School for Environmental Design and Rural Development, University of Guelph, Ontario, Canada. His primary areas of academic interest include the role of the state in development; environmental assessment and public participation; agricultural and environmental policy; and social movements.

Charis Enns is a Lecturer in International Development in the Department of Geography at the University of Sheffield and an Affiliated Researcher at the East African Institute at Aga Khan University. Her research focuses on the impacts that large-scale investments in land and natural resources have on rural landscapes and livelihoods in sub-Saharan Africa. She is also interested in corporate social responsibility and corporate-community engagement practices around sites of investment.

M. Omar Faruque is SSHRC Postdoctoral Fellow in the Department of Global Development Studies at Queen's University. Earlier, he was Assistant Professor of Sociology at the University of Dhaka and Research Associate in Agriculture and Rural Development Division at the Bangladesh Institute of Development Studies. His main areas of research interest are political sociology, development sociology, environmental sociology, and global & transnational sociology. He has published articles in *Social Movement Studies, Asian Journal of Political Science, The Extractive Industries and Society, Journal of Contemporary Asia, Bangladesh Unnayan Shamikkhaya*, and *Asian Journal of Social Science*. This chapter is based on his doctoral dissertation research on global capital, resource extraction, and political struggles in Bangladesh.

Grzegorz Foryś is sociologist and Professor at the Pedagogical University of Cracow (Institute of Political Sciences). In 1999 he graduated with first-class honours in sociology at Jagiellonian University in Cracow. At the same University he completed his PhD degree in 2006 and was awarded tenure in 2017. His scientific interests concentrate on social changes, social movements and rural sociology. He has recently been investigating the role of different social movements in promoting multi-functional rural development in Europe and especially in Poland. He is the author of *The Dynamic of Contention: Rural Protests in Poland 1989–2001* (Scholar 2008); *Agritourism Farms and Associations in Poland: In search of a Social Movement* (Scholar 2016); and *Politics: Society and the Economy in Contemporary Poland. An Introduction* (Scholar 2016). He is a member of the Polish Sociological Association and the European Society for Rural Sociology.

Ricardo Fuentealba is a PhD Candidate based at the Department of Human Geography, Planning, and International Development of the University of Amsterdam. His PhD project centres on disaster governance and urban equity impacts in Chile. Previously he was Adjunct Researcher at Rimisp – Latin American Center for Rural Development, where he worked on projects related to indigenous development, social movements, and regional development.

Kajsa Johansson is currently a PhD candidate at the institution for social science at Linnaeus University, Sweden. Her research examines the conditions for peasants' collective action in Mozambique since the country's Independence in 1975. Before initiating her PhD, she was working as a development practitioner in Mozambique, mainly with land rights, natural resources and agriculture policy in civil society support programmes.

Ahna Kruzic is a Food First/Institute for Food and Development Policy Fellow. Ahna's areas of interest include farm and rural justice, the intersections of gender and capitalist agricultural production, and the dismantling of white supremacy in food movement spaces.

Rebecca McEvoy is a MSc graduate in Rural Planning and Development from the University of Guelph, Ontario, Canada. Her research has centred around participatory land use planning, asset-based community development, and the impacts of community-driven opposition to land development proposals. She currently bridges theory and practice through her work in the non-profit sector as a land use planner, coordinating the rezoning and development of properties in Southwestern Ontario for affordable housing projects.

Tuija Mononen (DocSocSci, PhilLic) is working as a Senior Researcher in the Department of Geographical and Historical Studies, University of Eastern Finland. She has been involved in food research in social sciences since 1994 and mining research since 2009, and is a member of the societal mining research group in UEF (www2.uef.fi/en/kaivostutkimus/kaivos-ja-yhteiskunta). Her academic interest covers food issues, especially development, meanings and actors of organic production, food security and rural research and actor networks. In mining research, her interests cover societal questions of mining industry, community experiences and impacts of mining. In addition to refereed articles, she has edited books about food and mining issues.

Mariela Ramírez is Researcher at Rimisp – Latin American Center for Rural Development. Her work focuses on knowledge management for territorial development and environmental governance. She has collaborated in the design and implementation of public policies in the field of climate change adaptation and family farming.

Truong Quang Hoang is the Director of the Centre for Rural Development in Central Vietnam, a unit of the University of Agriculture and Forestry of Hue University, Vietnam. He is a university lecturer, a researcher, and also a practitioner working for many rural development projects and making links between research and development activities. He is interested in natural resources management, rural livelihood and community organizations, particularly looking for innovations from community, national and international levels for sustainable rural development.

Fumikazu Ubukata is a Professor of Environmental Policy and Economics at the Graduate School of Environmental and Life Science, Okayama University, Japan. He has a particular interest in development and environment in Southeast Asian countries, especially in Thailand, Laos, and Vietnam. His research mainly focuses on transformation of resource industries, institutional changes of resources management, commodification of natural resource and labour, and market-driven environmental management schemes.

Preface

Natural resource development takes multiple forms: deforestation and creation of plantations, dams, mines, pipelines, oil and gas drilling, fracking. Globalizing agricultural commodity markets and free trade agreements lead to land grabbing and agricultural land conversions. All these initiatives are driven by economic valuations while social and environmental effects are often given limited consideration. Projects result in negative environmental and social impacts and sometimes in population dislocations. But resistance is common. Rural social movements seek to defend environmental and social values. Projects are questioned and sometimes blocked. Even when projects continue, they are often affected by social movement efforts to reduce environmental and social damage.

The papers in this volume explore questions concerning rural social movement strategies and effects in relation to natural resource developments. Topics include land grabbing, development and energy corridors, mines, dams, pipelines, and resistance to agricultural industrialization. While most cases describe resistance, there are two that describe alternatives: compensation for environmental services and development of agritourism. The case studies consider movement organization, resource mobilization, action repertoires, alliance building, and framing processes. All describe the conditions leading to the emergence of the movements and the processes that unfolded as the movements resisted or sought to improve conditions. Some also consider short and longer-term movement outcomes asking whether and in what ways the movement was effective even if its primary goal of preventing a project was not achieved. Each case situates the movement within its broader social and political context. The cases provide examples of social movements contesting natural resource development in Africa, Asia, Latin America, North America, and Europe. Taken together, they provide a rich overview of current movements engaged in resisting the neoliberal agenda of global resource exploitation.

The papers collected in this volume were prepared for the International Rural Sociology Congress held at Ryerson University in Toronto in August 2016. The session on Social Movements Contesting Natural Resource

Development invited papers on natural resource based rural development under the dominant neoliberal paradigm. Preference was given to papers which presented detailed case studies and papers from the global South were particularly encouraged. The volume offers eight of the seventeen papers presented at the session, plus one that was prepared for the session but not delivered owing to logistical constraints.

Acknowledgements

I would like to thank the organizers of the International Rural Sociology Congress for creating the venue for the discussion of rural social movement issues. The session was initiated through the Rural Policy Learning Commons (RPLC) Natural Resources Development Network. The RPLC is an international partnership project funded by the Canadian Social Sciences and Humanities Research Council. I thank all the authors for their efforts in conducting the research, presenting, and writing up the final papers. I thank Tim Hardwick at Routledge for suggesting the volume and Amy Johnston of the Routledge Environment and Sustainability programme for nudging the process along through numerous delays. Most importantly, I would like to thank the thousands of active citizens who have raised their voices to contest natural resource developments that threaten the environmental and social systems upon which their livelihoods are based. They seek to prevent harm and to build more effective alternatives as the neoliberal juggernaut and escalating consumption demands erode our global sustainability.

Abbreviations

AEC	Asia Energy Corporation (Bangladesh) Pty Ltd
APC	Acción por los Cisnes (Action for the swans), Chile
APTN	Aboriginal Peoples' Television Network, Canada
ARA	Aggregate Resources Act, Ontario, Canada
BBC	British Broadcasting Corporation
BBS	Bangladesh Bureau of Statistics
BCP	Biocultural Community Protocol, Kenya
BHP	BHP Billiton Limited
BPRC	Bakken Pipeline Resistance Coalition, Iowa, USA
CBC	Canadian Broadcasting Corporation
CCC	Canadian Chefs' Congress
CDE	Consejo de Defensa del Estado (Defense Council of the State), Chile
CG	Caretaker Government, Bangladesh
CPP	Committee to Protect Phulbari, Bangladesh
CSR	Corporate Social Responsibility
CTV	CTV News Channel owned by Bell Media, Canada
DAE	Department of Agricultural Extension, Bangladesh
DFID	Department for International Development, United Kingdom
DSF	David Suzuki Foundation
EA	Environmental Assessment
EAA	Environmental Assessment Act
ECEAT Polska	European Centre for Ecological and Agricultural Tourism Poland
EIA	Environmental Impact Assessment
EMRD	Energy and Mineral Resources Division, Bangladesh
ENBSF	Ewaso Ng'iro Basin Stakeholder Forum, Kenya
F&WW	Food and Water Watch, USA
FES	Forest Environmental Services
FGI	focus group interview
FIAN	FoodFirst Information and Action Network
FMB	Forest Management Board

FPDF	Forest Protection and Development Fund, Vietnam
GCM	GCM Resources plc., London
GIS	Geographic Information Systems
GOB	Government of Bangladesh
GSFF	Global Solidarity Forest Fund, Mozambique
GTA	Greater Toronto Area, Canada
ICCI	Iowa Citizens for Community Improvement, USA
ICMM	International Council on Minerals and Metals
IDI	in-depth interview
IMF	International Monetary Fund
ISU SASA	Iowa State University's Sustainable Agriculture Student Association, USA
IWGIA	International Working Group on Indigenous Affairs
KaiELY	Kainuu Center for Economic Development, Transport and Environment, Finland
LAPSSET	Lamu Port–South Sudan–Ethiopia Transport corridor, Kenya
LCDA	LAPSSET Corridor Development Authority, Kenya
LLP	Limited Liability Partnership
MNR	Ministry of Natural Resources, Ontario, Canada
MOE	Ministry of the Environment, Ontario, Canada
NCBD	National Committee to Protect Oil, Gas, Mineral Resource, Power and Ports, Bangladesh
NDACT	North Dufferin Agricultural and Community Taskforce, Ontario, Canada
NGO	Non-Governmental Organization
NIMBY	Not-In-My-Backyard
NLC	National Land Commission, Kenya
NRD	Natural Resource Development
NRT	Northern Rangelands Trust, Kenya
NVCA	Nottawasaga Valley Conservation Authority, Ontario, Canada
OECD	Organization for Economic Co-operation and Development
OFA	Ontario Federation of Agriculture, Canada
ORAM	Organizacão Rural de Ajuda Mutua (Rural Mutual Aid Organization), Mozambique
PEPP	Petroleum Exploration Promotion Project, Bangladesh
PES	Payment for Environmental Services
PFES	Payment for Forest Environmental Services
PFTW	Polska Federacja Turystyki Wiejskiej (Polish Rural Tourism Federation)
PHARE	Poland and Hungary: Assistance for Restructuring Economies Program
PLN	Polish Zloty
PMO	Prime Minister's Office

PPC	Provincial People's Committee, Vietnam
PSC	Production Sharing Contract, Bangladesh
ROADS	Environmental Organization, Mozambique
RoK	Republic of Kenya
SAROS	State of the Aggregate Resource in Ontario Study
SEA	Strategic Environmental Assessment
SFE	State Forest Enterprise, Vietnam
SLU	Swedish University of Agricultural Sciences
UK	United Kingdom
UN	United Nations
UNAC	União Nacional de Camponeses (National Peasants Union) Mozambique
UNGPs	United Nations Guiding Principles on Business and Human Rights
UNOHCHR	United Nations Office of the High Commissioner for Human Rights
UPCN	União Provincial dos Camponeses de Niassa (Peasant's Union of Niassa Province) Mozambique
US	United States
VFDP	Vietnam Forests and Deltas Program
WFAN	Women, Food and Agriculture Network, USA
YLE	Finnish Broadcasting Company

1 Introduction

Social movements and natural resources

John F. Devlin

Introduction

This introductory chapter surveys main lines of social movement (SM) analysis. The chapter first suggests a simple framework for organizing the analysis of SMs. It then explores how the literature addresses three questions: What leads to the emergence of SMs? How do SMs act? What outcomes do SMs generate? The presentation demonstrates that SM studies have built up a comprehensive appreciation of the complexity of SMs, their roots, processes, and outcomes (Lofland, 1993). The chapter then offers a short review of literature oriented specifically to natural resource development and finishes with a brief overview of the cases presented in the chapters that follow.

A framework: the SM arena

The term 'social movement' refers to two related social objects: first, sets of actors working together to resist or promote some form of social change and second, the social process, sequence of events, or collective actions in which such SM actors are involved. The terms SM actors and SM processes will be used to differentiate these when appropriate. Diani suggests that nearly all definitions of SMs share three criteria: actors, conflict, and collective identity: 'a network of informal interactions between a plurality of individuals, groups and/or organizations, engaged in a political or cultural conflict, on the basis of a shared collective identity' (1992, p. 6). Snow, Soule, and Kriesi define SMs as:

> Collectivities acting with some degree of organization and continuity outside of institutional or organizational channels for the purpose of challenging or defending extant authority, whether it is institutionally or culturally based, in the group, organization, society, culture, or world order of which they are a part.
>
> (2007, p. 11)

McCarthy and Zald offer an ideational definition: SMs are sets of 'opinions and beliefs in a population' representing a preference for change (McCarthy & Zald, 1977, p. 1217). Tarrow provides a political definition of SMs as 'sustained challenges to powerholders in the name of a disadvantaged population living under the jurisdiction or influence of those powerholders' (1996b, p. 874). Tilly (2004) views SMs as major vehicles for ordinary people's participation in politics. Some scholars even go so far as to characterize contemporary democracies as 'movement societies' (Meyer & Tarrow, 1998). Taken together these definitions suggest that SMs are cumulative goal-oriented efforts by groups of actors to pursue a shared goal or goals seeking to change political institutions or policy or social institutions, social attitudes, or social behaviour.

SMs are found at many scales. Global movements such as the anti-slavery, the anti-colonial, the anti-globalization, or the environmental movement constitute 'transnational advocacy networks' (Keck & Sikkink, 1998; Tarrow, 2005). More geographically delimited SMs can include movements in specific international regions such as Pan-Africanism as well as national, sub-national, or local movements resisting a specific project in a specific place. Given this broad diversity, it is helpful to identify some basic analytic categories that are scale-neutral. The framework suggested here parallels Ostrom's framework for institutional analysis and Weible and Sabatier's advocacy coalition framework (Ostrom, 2007, 2010; Weible & Sabatier, 2006). The framework recognizes an 'internal' and an 'external' space and identifies a set of relevant factors or variables of interest in both spaces. In Ostrom's framework the internal space is an 'action arena', which contains actors, events, and patterns of interaction (processes). Inside this arena individuals 'interact, exchange goods and services, solve problems, dominate one another, feel guilty, or fight' (Ostrom, 2005, p. 829). In Weible and Sabatier's advocacy coalition framework, an 'internal' policy subsystem is identified made up of contending advocacy coalitions (actors) mobilizing policy beliefs and resources and seeking to influence the policy decisions of government actors (events and processes) (2006). This can be understood as the SM arena within which allies and antagonists seek to achieve their goals.

The SM arena is embedded within a broader context that includes factors such as the basic distribution of natural resources; fundamental socio-cultural values; institutional structures; and system rules. The movement arena is situated within and can be influenced by these socio-cultural and historical conditions, and other macro-historical factors and events which can affect the dynamic interactions between movement actors and their protagonists. The SM arena can be affected by external system events such as changes in socio-economic conditions; changes in public opinion; changes in governing coalitions; policy decisions; or specific events. The external conditions can help to explain movement emergence, dynamics, and outcomes through time. The framework is systemic meaning that over time outputs from the SM arena can potentially alter conditions inside the arena as well as external conditions through feedback loops. The framework helps to sort out some of the key features of SM processes and can set the stage for descriptions and explanations of SM action and outcomes. The framework invites analysts to identify SM actors and their allies, SM antagonists and their allies, and contextual conditions that affect both.

The most common depiction of a SM arena is to identify the SM actors and antagonists: peasants versus lords, workers versus capitalists, religious reformers versus established church. For political movements the state often plays the role of movement antagonist as in Flam's collection discussing relations between antinuclear movements and states (1994). But lifestyle movements may see the public or some alternative cultural group holding a set of shared beliefs as the prime antagonist rather than the state. Hence environmentalists oppose the chemical industry or the fossil fuel industry or even public indifference to environmental values, vegans oppose the meat industry, community organizers oppose developers, the #MeToo movement opposes male aggression.

The framework invites three modes of analysis: 1) analysis of the internal characteristics of movement actors, coalitions, and networks; 2) analysis of antagonistic engagements and contention between movement actors and their antagonists; and 3) contextual dynamics that influence and are influenced by SM actors and antagonists. This 'internal-external' framework is consistent with a wide range of SM analysis discussing the dynamics of SM actors and processes, the engagement of SM actors and antagonists, and the relationship between SM actors, SM processes, and the wider contexts within which SM processes unfold. These relationships cover the bulk of the analysis presented in social movement studies (McAdam, Tarrow, & Tilly, 2001).

This SM framework is applicable to the analysis of a single SM event, a sequence of events, a movement cycle, multiple cycles, or a movement history. The framework is also applicable to the comparative analysis of multiple movements and the formulation of generalizations perhaps leading to the identification of mechanisms and theories of how elements of movements interact.

Most case studies discuss all three elements of the framework to provide a deep description and analysis. Those seeking to generalize across social movements focus on more limited questions and sets of variables. Much of the richness of the literature arises from the capacity to bring a wide range of theory to bear on these more specific relationships. The framework is helpful in organizing efforts to answer three key questions: Why do movements emerge? How do they act? What outcomes do they generate?

SM emergence

Grievance has long been recognized as a source of SM emergence. Historically political economy has looked to material interests as an explanation of SM emergence. Marxists identified economic exploitation as the source of slave rebellions, peasant, and labour movements. Economic exploitation and capitalism continue to inform SM scholarship (Barker, Cox, Krinsky, & Nilsen, 2013; Della Porta, 2015). For Polanyi, SMs emerged as efforts to protect society from market induced disruptions arising from commodification of land and labour (Císař & Navrátil, 2017; Polanyi, 1957).

American sociology initially looked upon SM grievance as a symptom of social irrationality, pathology, or deviance. This approach was rooted in functionalist sociology and explained social movements as failures of social integration (Leites & Wolf, 1970; Tilly, Tilly, & Tilly, 1975; Turner & Killian, 1957). As Crossley argues recognizing grievances can render SM actions 'intelligible, understandable and thus rational' (2003, p. 289). But this functionalist approach was eclipsed by the observation that grievances may exist for many years without leading to the emergence of SMs. Racism in the United States existed for many years before the civil rights movement emerged (McAdam, 1982; McAdam, McCarthy, & Zald, 1996). The Arab Spring gave expression to long suppressed dissatisfaction with governments in the Middle East. Hence, grievances are not sufficient. They must be mobilized.

The resource mobilization approach sought to identify additional factors that transformed grievances into movements. Mobilization is stimulated by changes in available resources, group organization, and opportunities for collective action (Jenkins, 1983; McCarthy & Zald, 1977; Oberschall, 1973; Tilly, 1978). Marxist literature also identified organizational resources such as vanguards, vanguard parties, and organic intellectuals as instruments for the mobilization of economic grievances (Reynolds, 1992).

Marxist and resource mobilization analysts focused on movements for political change, but Touraine in the 1950s observed that SMs can emerge in response to hopes and aspirations that are not economically or materially determined and not oriented primarily to political change (Touraine, 1974, 2002). This line of analysis eventually came to be associated with 'new' social movements. Issues concerning identity and rights such as gender, sexuality, race, environment, ethnicity, disability, physical and mental health, HIV/AIDs, anti-poverty, and anti-surveillance can be understood as challenging established institutions or the dominant 'codes' of society (Melucci, 1980). Fuentes and Frank (1989) suggested that new SMs focus on morality and (in)justice and confront social power through mobilization against deprivation and for survival and identity. New SMs are often oriented towards changing ideas and attitudes of the broad public. They focus on changing society not necessarily by changing the state or policy.

Aberle (1966) suggested a SM typology built from two variables: 1) who the social movement seeks to change (everyone or specific individuals) and 2) how much change is being sought (limited or radical). This generated four types of social movements: alternative, reformative, redemptive, and revolutionary none of which were necessarily oriented towards material interests. Rucht and Neidhardt (2002) distinguish political change movements from personal change movements and both can be stimulated by non-material values.

The recognition of new SMs widened the scholarly universe of SMs (Crossley, 2002). But by the 1990s, social constructionist and discursive orientations made ideational factors, interpretation, and symbolization important dimensions of most SM analysis. Constructivism suggests that collective action is generated by actors assigning values to events, negotiating the meanings of those events and making decisions seeking to shape event outcomes (Snow & Oliver, 1995). The distinction between old and new SMs is thus less helpful since it is clear that all movements require an ideational foundation, and the distinction between values that are 'material' and values that are not, or goals that are political or not political are difficult to draw in practice. Suffragettes demanded the vote as well as recognitions of their social rights. The anti-smoking movement both wants the public at large to stop smoking and wants legislation to prohibit smoking in specific locations. At the same time changes in policy may require changes in lifestyle. Climate change policy will require changes in consumer behaviour. The politics–lifestyle boundary is very permeable.

Emergence as a phase in a SM history is not easily distinguished from the actions that a SM undertakes. Ideas may emerge among a few individuals long before the ideas have attracted sufficient attention to be considered a SM. SMs build through time. It is thus helpful to think of emergence not simply as a temporal question about what happens early in the life of a SM, but as a question about foundational motivations which both create and continually support the reproduction and growth of the movement. But for an emergent SM to persist through time and grow, ideas and values are not sufficient. There must be action. The analysis of SM action draws the focus towards SM processes that have been a central focus of SM studies.

SM processes

Investigations of SM process is a literature rich in description, concepts, and theory. Much recent work suggest that one should begin any general overview of movement processes by seeking to understand how the ideas and values that are foundational to the emergence of the movement have been mobilized through action. This work falls under the concept of 'framing'.

Framing

Goffman (1974) drew attention to the ways in which information is interpreted or misinterpreted and how this might impact SM participation. A movement's interpretive framework expresses what is important to the participants in the movement and suggests an agenda for change. Meaning is constructed as different events, experiences, facts, and values are emphasized or downplayed. Whatever the foundational impetus for SM emergence, framing is crucial for movement reproduction and growth. A given social issue may not be of interest to the general public until it has been framed in a way generating concern for a wider audience (Dewulf, Craps, & Dercon, 2004). Framing affects how issues are understood by SM participants and can be part of how the SM presents itself to new potential sympathizers.

Frames such as justice; rights; equity; freedom; independence; identity; history; livelihood; truth; sustainability; or peace are central appeals that attract participants and sustain their participation. They also can serve to build sympathy among a wider public who constitute part of the external SM environment. In relation to movement participants and potential participants, frames are shaped with the intention of rallying public support and engaging bystanders. Through framing activity, a SM can align similar frames, bridge frames that are ideologically compatible and yet structurally distinct, amplify frames to increase the strength of the frame as a source of movement attention, and extend frames to expand their boundaries so that more people are attracted to and drawn into the movement. Frame transformation involves a change in the way that a situation or idea is fundamentally understood, such that what was once considered a non-issue or at the very least, tolerable, is now seen as an injustice that must be righted (Benford & Snow, 2000; Gamson, 1999; Gamson, Fireman, & Rytina, 1982; Snow, Rochford Jr, Worden, & Benford, 1986). Framing uses ideas and language that can resonate with the broader culture. This resonance is a legitimating device which increases buy-in. Snow and Benford (1988) call this 'narrative fidelity' (p. 210). Even among SM allies, there can be contention and multiple 'poles' around which actors congregate (Crossley, 2005). Effective movement building finds ways to construct frames that keep allies. Framing can support perceptions of grievance, blame, causality, popular support, strategy, and tactics (Benford, 1997; Snow & Benford, 1992). Frame alignment seeks to find shared beliefs and values with potential movement participants or by creating new shared beliefs and values among a broader constituency (Taylor, 2000; Zavestoski, Agnello, Mignano, & Darroch, 2004).

By the 1980s, framing was identified as bringing in an interpretive, ideational, cultural, or discursive dimension into movement studies. Snow, et al. (1986) laid out a typology of frame alignment processes. They identified five distinct types of claims that can spark SM participation. These include: claims about the gravity of the issue or grievance; claims about who is to blame for the issue; claims about the perceived intention or moral character of the antagonists; claims about the probability of effecting a change; and claims about the necessity of participation and taking action.

In relation to movement antagonists, frames serve to articulate grievance and undermine opposing viewpoints. The framing presents a call to action, challenges authority, and problematizes the status quo (Snow, Vliegenthart, & Corrigall-Brown, 2007). Devlin and Yap (2008) argue that framing is not only about values. It also has a technical or scientific dimension. Academics, engineers, and scientists can contribute to SMs by offering credible technical frames and challenging the frames of opponents. Challenging the technical assertions of opponents prevents SMs from being dismissed as a self-serving or local special-interest groups. The addition of technical expertise can lead to the provision of alternative policy and development scenarios for authorities to consider, as well as provide critiques and deconstruction of the assertions of antagonists.

Framing contributes to the creation of movement identities. The roots of this approach can be found in Melucci, Keane, & Mier's (1989) description of the construction of collective identities in the 'submerged networks' of everyday life, in which new, experimental worldviews and social relationships are developed by small groups in response to emergent tensions' (cited in Mische, 2003). Polletta and Jasper (2001) examine the role of identity in the creation of collective claims, recruitment into movements, strategic and tactical decision making, and movement outcomes. They suggest that collective identity has been often used to fill in gaps in structuralist, state-centred, and rationalist accounts of social movements.

Crossley (2005) emphasizes that conceiving of SMs as ideational fields with multiple interpretive poles (frames) helps explain how SMs change over time. Interaction among movement members from experienced activists to neophytes allows SM actors to learn and negotiate their overlapping but not completely shared interpretation of movement issues. SMs are not monolithic, new ideas and strategies emerge and are absorbed. As a result, movements evolve. Frames provide the foundational content for a SM but frames must be projected into the SM arena to attract adherents and challenge antagonists. SM frames are presented through a wide assortment of SM actions, what Tilly (2005) calls the SM repertoire.

Repertoires

Many SM actions are familiar forms of normal politics. SM actors provide interviews; organize and participate in focus groups; hold meetings and workshops; attend and speak at public meetings; conduct research; prepare and submit briefs and reports; call for and engage in public hearings; lobby decision makers; create special-purpose coalitions; collect signatures and present petitions; distribute pamphlets; participate in radio and television programmes; issue news releases; produce newsletters; buy advertising; generate education programmes; create websites; blogs; polls; participate in citizen advisory committees or boards; undertake legal actions; organize consumer boycotts; and stimulate cultural action through the creation of art, media, film, comedy, and internet campaigns. These elements of normal politics are also performed by interest groups not identified as SMs. The boundary between interest groups and SMs is soft. SMs have a larger and less specific engagement than interest groups, have no authoritative organizational centre, and no formal membership. Interest groups are often formally organized, claim a representative function, and have formal memberships (Wagner, Polak, & Świątkiewicz-Mośny, 2019). But SMs often engage in public spectacles: processions; vigils; rallies; boycotts; demonstrations; hunger strikes, which are efforts to call public attention to the movement's concerns. Most interest groups do not use such tactics. Tilly (2004) identifies such displays as efforts to communicate worthiness, unity, numbers, and commitments to the movement actors and/or their constituencies. These actions project the SM's frames into the SM arena.

The SM repertoire also includes actions that push beyond cultural and legal norms and engage in 'transgressive' actions: illegal marches and demonstrations; disruption of public meetings, land and building occupations, road barricades and blockades, spiking trees, chaining people to fences or buildings. At even more extreme levels SM actors may engage in large-scale disruption, political violence, bombings, terrorism, and military actions constituting insurrection and civil war. While the range of repertoires is continuous there is no necessity for escalation from one repertoire to another. SM actors choose their tactics and which tactics are chosen can be an important marker of cleavages even within a movement with shared goals (González, 2019). With so many strategic and tactical options decisions about how to act can be sources of cleavage among SM allies. Crossley (2005) suggests that even among allies the SM arena can be a 'field of contention' in which there may be multiple 'poles' around which allies congregate.

Attention to frames and repertoires captures much but certainly not all of the thematic range of analysis with respect to SM processes. Looked at chronologically attention to SM processes was first drawn towards major social conflicts such as peasant movements and labour movements. In American sociology, the lens of grievance gave way to a focus on resources and resource mobilization. It is always necessary for a SM to mobilize resources. The resources are organization, money, knowledge, and strategy. Resources are the means by which SMs can act. The resource mobilization approach argued that formally structured organization was more effective than informal and decentralized organization at mobilizing resources and movement success was determined by strategic interactions (Gamson, 1975; Jenkins, 1983). Marxists also promoted organization in the form of the vanguard party and centralized decision making. More recently, increased attention has been paid to decentralized or network organizational structures with low levels of institutionalization (Mische, 2003). Viewing SMs as networks emphasizes information as the central organizational resource and thus points back to framing as a foundation activity. Mische (2003) points to the important effects of bridging and overlapping across multiple networks to expand the movement's allies. McDonald (2002) argues that anti-globalization movements are fluid networks that cannot be explained in terms of a single collective identity or even a shared personalized commitment. The anti-globalization frame has attracted many actors with different focal points and sources of grievance (Caouette & Kapoor, 2015). Newer SMs such as Occupy Wall Street have rejected organization building altogether and rely completely on the strength of voluntary networks.

The concept of the political opportunity structure was an explicit recognition of how changes in external conditions may affect the actions of movement actors and the outcomes they achieve (Meyer & Staggenborg, 1996; Van Der Heijden, 2006). Tarrow (1996a) argues that changes in context can create new or expanded opportunities for effective action. These may be based in the dynamics of policy issues or the dynamics of specific groups. Geographically or temporally different state structures may also influence opportunities leading to the comparative or the historical study of opportunity structures. Historical explanations also place the focus on the contention between movement and non-movement actors while attending to external factors such as critical historical conjunctures or events (Paige, 1999).

SM outcomes

Goal achievement is a basic criterion for the assessment of movement success. A movement may seek to capture the state or change a government, it may seek to introduce or change a policy, or make fundamental changes in social behaviour. It may seek to start or stop a project or programme. If these outcomes are achieved, then the movement can be readily described as successful. But such clear-cut outcomes are rare. Turner and Killian suggested three criteria of SM success: benefits for members; changes in power relations; and the realization of a 'program for the reform of society' (1972, p. 256). Staggenborg (1995) recognized three similar categories of SM outcomes: political and policy outcomes, mobilization outcomes, and cultural outcomes. Political and policy outcomes include successful demands for changes in laws (Bernstein, 2003). Mobilization outcomes refer to changes in the internal capacity of the SM. If the SM grows then mobilization has been successful whether or not movement goals are achieved. Diani (1997) assesses SM outcomes in relation to their effectiveness in creating social capital. Cultural outcomes refer to changes in social attitudes and behaviour.

Andia and Chorev (2017) exemplify the potential diversity of SM outcomes through three cases of transnational lobbying at the World Health Organization. The tobacco campaign resulted in a binding treaty; the infant formula campaign resulted in a non-binding code of conduct; and the pharmaceutical campaign failed because no agreement could be reached. They suggest that SM success or failure depends upon a wide range of variables: resources mobilized – money, organization, allies, expertise, perceived legitimacy of the knowledge supporting the advocates' claims, and on the prestige of the experts participating in the campaigns (Andia & Chorev, 2017). To put this succinctly, SM success is contextual. No SM is guaranteed success.

Social movements and natural resources

The investigation of SMs relating specifically to natural resources has generated a wide diversity of descriptive and theoretical work consistent with the diverse elements of the review of concepts and issues offered earlier. The literature most commonly finds natural resource developments threatening communities and stimulating efforts to mobilize resistance often through calls for an end to development or a change in policy. Most analysts offer an assessment of emergence, process, and outcomes while focusing on specific dynamics within the SM arena and suggesting how the external SM environment has influenced events.

A foundational natural resource is land, and land appears recurrently in SM analysis beginning with peasant movements and land reform movements (Wolford, 2010). Edelman (2012) notes that the rural actors involved in social movements include 'smallholders, pastoralists, tenant farmers, sharecroppers, squatters, fisher folk, forest dwellers, indigenous peoples, landless labourers, large landowners and agro-industrial entrepreneurs' (p. 431). Bebbington (1996) examined the impact of agrarian commodification on Indigenous movements in Ecuador. Camus (2013) describes how peasant movements mobilizing against neoliberalism in Latin America have consolidated their framing around the issue of food sovereignty. Stedile (2002) explains how Brazil's peasant-based Movimento Sem Terra uses land occupations as a central strategy. Rosset and Martinez-Torres (2012) note how agroecology is being mobilized in resistance to agribusiness and corporate land-grabbers by peasant movements, in particular, La Via Campesina. Brown (2014) explores why Kheti Virasat Mission, an agricultural movement in Punjab, India chose to adopt a SM model of organization and to reject an NGO organizational form due to the concern for autonomy from donors and avoidance of a middle-class identity. McKeon (2013) analyses how rural social movements have built up their capacities as global mobilizers and policy players and exploiting the political opportunities created by the linked global food, fuel, climate, and financial crises with respect to land grabbing.

Deforestation has drawn attention particularly in relation to biodiversity decline, soil erosion, climate change, and cultural survival. Michaelson (1994) used his study of Wangari Mattai's Greenbelt Movement in Kenya to explore the potential for consensus movements. Mawdsley (1998) criticized neopopulist interpretations of the Chipko forest protection movement in Uttaranchal, India as a gender and environment movement, pointing out that this neopopulist interpretation missed important livelihood and regional identity threads that also stimulated the movement agenda. Zhouri (2004) suggested how global movements to challenge Amazonian deforestation may have contributed to a simplification of the complexity of Amazonian life and unintentionally perpetuated patterns of inequality. Bartley (2007) showed how grant-making foundations used the threat of protest to build support for forest certification programmes as an organizational field of moderate governance by creating a market-oriented alternative to disruptive environmental boycotts. Di Gregorio (2012) draws from the experiences of the Indonesian forest tenure movement three propositions about networking and coalition building: that movement networks tend to be most dense when the variety of environmentalism is shared; that framing activities influence the density of interaction among organizations; and that discourse coalitions can bridge across distinct environmental values. Greenleaf (2019) considers how land rights and deforestation in the Brazilian Amazon are being influenced by issues of carbon sequestration. Moore (2015) discusses the struggle against deforestation in Clayoquot Sound, Canada and how it helped to redefine the meaning of eco-feminism.

The extraction of fossil fuels is another important terrain for SM action. Osaghae (2008) examines uprisings by minority groups in the oil-rich Niger Delta region of Nigeria that have transformed struggles from 'accommodation-seeking nationalism to self-determination nationalism'. Curran (2017) suggests that the hope to create a social licence to operate has opened up space for opposition to use a democracy frame to challenge coal seam gas (hydraulic fracturing) development. Healy and Barry (2017) examine the fossil fuel divestment movement and the potential it presents for a just transition to renewable energy technologies.

Hydroelectric and irrigation dam developments have given rise to major SMs such as the Narmada Bachao Andolan (NBA), the save the Narmada movement. Routledge (2003) examines various repertoires of resistance employed by the NBA. Gandhi (2003) explores the complexity of the Narmada conflict which generated tension over afforestation in tribal villages, resettlement in regional centres, and transnational lobbying of donor agencies resulting in disarticulation between parties who shared a general orientation of resistance to the dam development. McCormick (2007) presents the anti-dam movement in Brazil as a science democratization movement that protests the use of science to justify projects that displace poor, rural people.

Mining is another important SM sector. Dahlberg-Grundberg and Örestig (2017) demonstrate how an anti-mining campaign in Kallak, Sweden, combined on-site resistance with social media strategies that successfully linked the local protest to the globally contentious mining boom and its impact of indigenous communities. Bebbington, Bebbington, Bury, Lingan, Muñoz, and Scurrah (2008) examine SMs confronting mining developments and their effects on livelihoods and rural territorial development in the Andes.

These are a just a few examples of the extensive and ever-growing literature focusing on SMs and natural resources. The cases presented in the following chapters add to this literature. They discuss land grabs, infrastructure corridors, mines, dams, resource processing plants, and pipelines. The authors discuss emergence, process and outcomes with respect to these natural resource development projects. The cases include examples of confrontation but also of positive SM outcomes with respect to agricultural tourism where movements have succeeded to promote and protect resources. The authors represented a range of theoretical or conceptual starting points. The chapters are presented geographically with two chapters discussing movements from each of the continents of Africa, Asia, North America, and Europe, and one chapter discussing Latin America.

The cases

In Chapter 2, Kajsa Johanssen examines the emergence over the past decade of collective action by peasants in Niassa Province in northern Mozambique. Since independence in 1975, peasant collective actions have been rare. She argues that the recent actions have been stimulated by the introduction of forest plantations. Although privatization of land is prevented by the Mozambican land law, the government has given land concessions to large plantation firms. This has resulted in loss of land access and limited employment opportunities for peasants. She explains the peasant reaction using Polanyi's concept of embeddedness. The loss of access to the plantation land has resulted in a struggle to preserve the peasant economy. The land concessions not only represent something foreign to the peasant logic, they actually threaten the possibility to continue life according to that logic. Peasant collective action in Niassa can be understood as a countermovement to the disembedding of land represented by government land concessions and plantation development.

In Chapter 3, Charis Enns and Brock Bersaglio examine the Lamu Port–South Sudan–Ethiopia Transport corridor which began construction in Kenya in 2012. Recent discoveries of oil, geothermal, and wind resources have attracted foreign investors to the region. The corridor plan includes a pipeline for transport of crude oil to the Indian Ocean, a new port facility, a new highway, railway, and other transport infrastructure. The corridor has been promoted as a catalyst for industrial and agricultural development and the expansion of social and economic services benefiting rural groups in northern Kenya. The chapter considers how this state-led process of territorial restructuring is impacting two rural groups: pastoralists and conservationists, and how they are responding. Enns and Bersaglio argue that these two groups are engaging in acts of (re)territorialization 'from below' to resist territorial restructuring 'from above'. The chapter argues that these efforts are having uneven outcomes, aggravating historical anxieties, and generating new points of contention relating to their different relationships to the land.

In Chapter 4, M. Omar Faruque examines the decade-long political struggle against a British mining corporation in Bangladesh, GCM Resources Plc, and its wholly-owned subsidiary, Asia Energy Corporation (Bangladesh) Pty Ltd. Locally, there was resistance to the construction of a large open pit coal mine in Phulbari, a rural town 300 kilometres northwest of Dhaka. Nationally, there was pressure for restructuring coal policy. The chapter considers mobilization strategies and movement outcomes to show how these local and national movements persuaded powerful authorities to block the mine and change national policy. Key features of the successful movement include a focus on the underestimation by 'corporate science' of the social, economic, environmental, and political risks associated with mining; the use of the 'politics of time' as well as the 'politics of scale'; the critical role of local community leaders; and the engagement of a radical nationalist movement that could control messages and resources. In the end, the mining company could not meet the challenge of these opposition forces and the government yielded to the pressure.

Chapter 5 by Fumikazu Ubukata and Truong Quang Hoang examines the processes and local responses to a programme called Payment for Forest Environmental Services (PFES) in Central Vietnam's Thua Thien Hue province. The authors briefly introduce the PFES programme in Vietnam, describe how it has been implemented in Thua Thien Hue province, describe the local response in a village resettled due to dam construction, and conclude with a discussion of how different meanings of nature are practiced and negotiated in this authoritarian socio-political setting. The argue that the state dominated how forest environmental services (FES) were defined and commoditized. The state created the 'market' and 'scientified' FES to strengthen its ability to govern the PFES chain. The state's domination of the PFES programme caused confusion in actual local implementation and also precluded direct negotiation between villagers and hydropower companies. The introduction of PFES in the study area did not create a sense of fairness in natural resource development. When villagers found that petitions and negotiations failed, they conducted everyday resistance through illegal encroachment or shirking of conservation tasks.

In Chapter 6, Ricardo Fuentealba and Mariela Ramírez discuss the citizen-based movement which emerged in Valdivia, southern Chile, in response to the death of black swans in 2004 and which gained national and international attention. The deaths were due to effluent from a cellulose plant 56 kilometres north of the city. The plant belonged to the CELCO-Arauco company, the world's second largest cellulose producer. The movement for the swans was self-funded, self-organized, and had strong local roots. It led actions against the plant and against the state for its passive role during the disaster. The chapter analyses the characteristics of the socio-environmental movement in Valdivia, the immediate and longer-term political impacts related to this movement, and the mechanisms through which these impacts were generated. Although the cellulose plant was not eliminated and thus the primary objective of the movement was not achieved, the movement led to positive environmental transformations, encouraging more democratic forms of decision-making and resource governance by making visible the lack of transparency and community involvement in environmental assessment of private projects in Chile. These wider impacts are explained by both, the movement's organization and its strategies, and the broader context in which the mobilization was developed. Fuentealba and Ramírez conclude that social movements that fail to stop specific natural resource development projects can still contribute to broader positive environmental transformations over the long term encouraging more democratic forms of decision-making and resource governance.

In Chapter 7, Angie Carter and Ahna Kruzic describe the Bakken Pipeline Resistance Coalition, an Iowa-based grassroots group that began organizing against the Dakota Access pipeline in late summer 2014. They analyse how the Coalition generated a new environmental resistance narrative in the state which could be shared by previously antagonistic parties who were united against the pipeline. Dakota Access completed construction of the controversial pipeline in the summer of 2016 and oil began flowing in spring of 2017, yet the Coalition continues to organize grassroots resistance. These efforts successfully reframe previously polarizing narratives about agricultural production and pollution into a more inclusive, broader-reaching narrative centring on the commons and uniting pipeline opponents from across existing rural/urban and political divides. They identify three strategic processes used by the Coalition to create cultural narrative shifts: creating space for new claims-makers, elevating the commons, and aligning values with regional and national campaigns against extractive energy. They argue that the Coalition's intentionality in creating space for a multitude of voices and a variety of perspectives helped the movement to grow beyond the landowners with farms in the pipeline's path. These new voices reframed a defensive narrative about Iowa's farmland and created a narrative of protection. The Coalition also aligned Iowans' anti-pipeline struggle to extractive energy campaigns beyond Iowa through an emphasis on the importance of neighbouring. Together, these processes created a narrative-shifting arena that transformed the symbolic landscape.

Agricultural land is the focal resource in Chapter 8, where Rebecca McEvoy and John F. Devlin describe the community mobilization opposing a controversial open-pit quarry proposal for mining amabel dolostone bedrock in Melancthon Township, a rich agricultural region in Ontario, Canada. Given the policy context in Ontario where quarries are not subject to environmental assessment and where many previous contentious quarry applications have been approved, the development of the Melancthon open-pit quarry seemed likely. This chapter examines the discursive framing presented by the contending sides and assesses how framing focused on food led to the eventual success of the opposition movement when after one and one-half years the quarry proposal was withdrawn by the proponent. The chapter explores how farmers, landowners, environmental groups, academics, other technical specialists, and celebrities strengthened the work of the community-based social movement.

In Chapter 9, Tuija Mononen and Ismo Björn focus on three Finnish mining cases. They demonstrate that mining activism has existed for a long time in Finland with a particular focus on water quality issues. The main repertoires of activism have included meetings, letters, and requests to the authorities and other stakeholders. In earlier cases activism remained local because the mines were remote and mining companies were able to manage the publicity. But the mining boom of the 2000s brought mining issues to national attention. Now, non-local activist groups have entered the mining arena with a wider capacity to generate a national movement using social media. They also observe that anti-mining movements in Finland do not easily fit into either the 'old' or the 'new' social movement categories. Some appear old by presenting an anti-business agenda seeking to close down mines and stop all mining in Finland. However, movement motives are not strictly economic. Environmental and quality of life issues are central thus fitting the model of new social movements. The authors conclude that Finnish mining activism is multifaceted and rural people have shown over time that they are not willing to accept any and all possibilities of local and regional economic benefits at the cost of a decline in environmental quality.

In Chapter 10, Grzegorz Foryś presents the agritourism movement in Poland as a new social movement. He argues that transformation processes in rural Poland combined with EU and state policies aimed at supporting the diversification of agriculture and the income sources of rural residents are encouraging the establishment of agritourism farms among a relatively large group of small farm owners who are weakly attached to the agricultural market. The agritourism movement focuses on the numerous natural resources still present in Polish rural areas and is supported by a cultural community oriented to the preservation of post-material traditional and ecological values. However, he warns that the image of the countryside upon which the movement is built may be undermined by increased commodification of rural resources through the development of rural consumption activities and this is one of the main developmental dilemmas that Polish rural areas will face in the future.

These chapters present detailed case studies that demonstrate the diversity of struggles stimulated by natural resource development. They describe victories and defeats, immediate and longer-term effects, repertoires of action, political and cultural work. All have emerged from a concern with the impact or potential impact of resource developments in specific places. They demonstrate a wide range of SM actions and a diversity of SM outcomes. Taken together, they represent SMs at work across the globe seeking to avoid the potential harm natural resource developments can have to the environment and the communities that are affected by them.

References

Aberle, D. F. (1966). *The Peyote Religion among the Navaho*. Chicago, IL: Aldine.

Andia, T. & Chorev, N. (2017). Making knowledge legitimate: transnational advocacy networks' campaigns against tobacco, infant formula and pharmaceuticals. *Global Networks*, 17, 2, 255–280.

Barker, C., Cox, L., Krinsky, J., & Nilsen, A. G. (Eds). (2013). *Marxism and Social Movements*. Leiden: Brill.

Bartley, T. (2007). How foundations shape social movements: The construction of an organizational field and the rise of forest certification. *Social Problems*, 54, 3, 229–255.

Bebbington, A. J. (1996). Movements, modernizations, and markets: Indigenous organizations and agrarian strategies in Ecuador. In R. Peet & M. Watts (Eds), *Liberation Ecologies: Environment, Development, Social Movements*. London: Routledge.

Bebbington, A., Bebbington, D. H., Bury, J., Lingan, J., Muñoz, J. P., & Scurrah, M. (2008). Mining and social movements: struggles over livelihood and rural territorial development in the Andes. *World Development*, 36, 12, 2888–2905.

Benford, R. D. (1997). An insider's critique of the social movement framing perspective. *Sociological Inquiry*, 67, 4, 409–430.

Benford, R. D. & Snow, D. A. (2000). Framing processes and social movements: An overview and assessment. *Annual Review of Sociology*, 26, 1, 611–639.

Bernstein, M. (2003). Nothing ventured, nothing gained? Conceptualizing social movement 'success' in the lesbian and gay movement. *Sociological Perspectives*, 46, 3, 353–379.

Brown, T. (2014). Negotiating the NGO/social movement dichotomy: Evidence from Punjab, India. *VOLUNTAS: International Journal of Voluntary and Non-profit Organizations*, 25, 1, 46–66.

Caouette, D. & Kapoor, D. (Eds). (2015). *Beyond Colonialism, Development and Globalization: Social Movements and Critical Perspectives*. London: Zed Books.

Císař, O. & Navrátil, J. (2017). Polanyi, political-economic opportunity structure and protest: Capitalism and contention in the post-communist Czech Republic. *Social Movement Studies*, 16, 1, 82–100.

Crossley, N. (2002). *Making Sense of Social Movements*. Philadelphia, PA: Open University Press.

Crossley, N. (2003). Even newer social movements? Anti-corporate protests, capitalist crises and the remoralization of society. *Organization*, 10, 2, 287–305.

Crossley, N. (2005). How social movements move: From first to second wave developments in the UK field of psychiatric contention. *Social Movement Studies*, 4, 1, 21–48.

Curran, G. (2017). Social licence, corporate social responsibility and coal seam gas: framing the new political dynamics of contestation. *Energy Policy*, 101, February, 427–435.

Dahlberg-Grundberg, M. & Örestig, J. (2017). Extending the local: activist types and forms of social media use in the case of an anti-mining struggle. *Social Movement Studies*, 16, 3, 309–322.

Della Porta, D. (2015). *Social Movements in Times of Austerity: Bringing Capitalism Back into Protest Analysis*. Cambridge: Polity Press.

Devlin, J. F. & Yap, N. T. (2008). Contentious politics in environmental assessment: blocked projects and winning coalitions. *Impact Assessment and Project Appraisal*, 26, 1, 17–27.

Dewulf, A., Craps, M., & Dercon, G. (2004). How issues get framed and reframed when different communities meet: A multi-level analysis of a collaborative soil conservation initiative in the Ecuadorian Andes. *Journal of Community & Applied Social Psychology*, 14, 3, 177–192.

Di Gregorio, M. (2012). Networking in environmental movement organisation coalitions: Interest, values or discourse? *Environmental Politics*, 21, 1, 1–25.

Diani, M. (1992). The concept of social movement. *The Sociological Review*, 40, 1, 1–25.

Diani, M. (1997). Social movements and social capital: A network perspective on movement outcomes. *Mobilization: An International Quarterly*, 2, 129–147.

Edelman, M. (2012). Rural social movements. In E. Amenta, K. Nash, & A. Scott (Eds). *The Wiley-Blackwell Companion to Political Sociology* (pp. 431–443). Oxford: John Wiley & Sons.

Flam, H. (1994). A theoretical framework for the study of encounters between states and anti-nuclear movements. In H. Flam, (Ed.). *States and Anti-Nuclear Movements* (pp. 8–26). Edinburgh: Edinburgh University Press.

Fuentes, M. & Frank, A. G. (1989). Ten theses on social movements. *World Development*, 17, 2, 179–191.

Gamson, W. A. (1975). *The Strategy of Social Protest*. Homewood, IL: Dorsey Press.

Gamson, W. A. (1999). The success of the unruly. In D. McAdam & D. A. Snow (Eds). *Social Movements: Readings on Their Emergence, Mobilization, and Dynamics* (pp. 357–364). Los Angeles, CA: Roxbury.

Gamson, W. A., Fireman, B., & Rytina, S. (1982). *Encounters with Unjust Authority*. Homewood, IL: Dorsey Press.

Gandhi, A. (2003). Developing compliance and resistance: The state, transnational social movements and tribal peoples contesting India's Narmada Project. *Global Networks*, 3, 4, 481–495.

Goffman, E. (1974). *Frame analysis: An essay on the organization of experience*. Cambridge, MA: Harvard University Press.

González, R. (2019). From the squatters' movement to housing activism in Spain: Identities, tactics and political orientation. In N. M. Yip, M. A. M. López, & X. Sun, (Eds) *Contested Cities and Urban Activism* (pp. 175–197). Singapore: Palgrave Macmillan.

Greenleaf, M. (2019). The value of the untenured forest: land rights, green labor, and forest carbon in the Brazilian Amazon. *The Journal of Peasant Studies*, 1–20. Online. DOI: 10.1080/03066150.2019.1579197.

Healy, N. & Barry, J. (2017). Politicizing energy justice and energy system transitions: Fossil fuel divestment and a 'just transition'. *Energy Policy*, 108, 451–459. Online. https://doi.org/10.1016/j.enpol.2017.06.014.

Jenkins, J. C. (1983). Resource mobilization theory and the study of social movements. *Annual Review of Sociology*, 9, 527–553.

Keck, M. & Sikkink, K. (1998). *Activists Beyond Borders*. Ithaca, NY: Cornell University Press.

Leites, N. & Wolf Jr, C. (1970). *Rebellion and Authority: An Analytic Essay on Insurgent Conflicts*. Santa Monica, CA: Rand Corp. (No. RAND-R-462-ARPA).

Lofland, J. (1993). Theory-bashing and answer-improving in the study of social movements. *The American Sociologist*, 24, 2, 37–58.

Mawdsley, E. (1998). After Chipko: From environment to region in Uttaranchal. *Journal of Peasant Studies*, 25, 4, 36–54.

McAdam, D. (1982). *Political Process and the Development of Black Insurgency*. Chicago, IL: University of Chicago Press.

McAdam, D., McCarthy, J. D., & Zald, M. N. (Eds) (1996). *Comparative Perspectives on Social Movements: Political Opportunities, Mobilizing Structures and Cultural Framings*. Cambridge: Cambridge University Press.

McAdam, D., Tarrow, S., & Tilly, C. (2001). *Dynamics of Contention*. Cambridge, MA: Cambridge University Press.

McCarthy, J. & Zald, M. (1977). Resource mobilisation and social movements: A partial theory, *American Journal of Sociology*, 82, 6, 1212–1241.

McCormick, S. (2007). Democratizing science movements A new framework for mobilization and contestation. *Social Studies of Science*, 37, 4, 609–623.

McDonald, K. (2002). From solidarity to fluidity: Social movements beyond 'collective identity' – the case of globalization conflicts. *Social Movement Studies*, 1, 2, 109–128.

McKeon, N. (2013). 'One does not sell the land upon which the people walk': Land grabbing, transnational rural social movements, and global governance. *Globalizations*, 10, 1, 105–122.

Melucci, A. (1980). The new social movements: A theoretical approach. *Information (International Social Science Council)*, 19, 2, 199–226.

Melucci, A., Keane, J., & Mier, P. (1989). *Nomads of the present: Social movements and individual needs in contemporary society*. Philadelphia, PA: Temple University Press.

Meyer, D. S. & Staggenborg, S. (1996). Movements, countermovements, and the structure of political opportunity. *American Journal of Sociology*, 101, 6, 1628–1660.

Meyer, D. S. & Tarrow, S. (1998). A movement society: Contentious politics for a new century. In D. S. Meyer & S. Tarrow, (Eds) *The Social Movement Society: Contentious Politics for a New Century* (pp. 1–28). New York: Rowman & Littlefield.

Michaelson, M. (1994). Wangari Maathai and Kenya's Green Belt Movement: Exploring the evolution and potentialities of consensus movement mobilization. *Social Problems*, 41, 4, 540–561.

Mische, A. (2003). Cross-talk in movements: Reconceiving the culture-network link. In M. Diani & D. McAdam (Eds) *Social Movements and Networks: Relational Approaches to Collective Action* (pp. 258–279). Oxford: Oxford University Press.

Moore, N. (2015). *The Changing Nature of Eco/feminism: Telling Stories from Clayoquot Sound.* Vancouver: UBC Press.

Oberschall, A. (1973). Social conflict and social movements. Englewood Cliffs., NJ, Prentice-Hall.

Osaghae, E. E. (2008). Social movements and rights claims: The case of action groups in the Niger Delta of Nigeria. *VOLUNTAS: International Journal of Voluntary and Nonprofit Organizations*, 19, 2, 189–210.

Ostrom, E. (2005). Doing institutional analysis digging deeper than markets and hierarchies. In C. Ménard & M. M. Shirley (Eds) *Handbook of New Institutional Economics* (pp. 819–848). Boston, MA: Springer.

Ostrom, E. (2007). Institutional rational choice: An assessment of the institutional analysis and development framework. In P. A. Sabatier (Ed.) *Theories of the Policy Process, Second Edition* (pp. 21–64). Boulder, CO: Westview Press.

Ostrom, E. (2010). Beyond markets and states: Polycentric governance of complex economic systems. *The American Economic Review*, 100, 3, 641–672.

Paige, J. M. (1999). Conjuncture, comparison, and conditional theory in macrosocial inquiry. *American Journal of Sociology*, 105, 3, 781–800.

Polanyi, K. (1957). *The Great Transformation.* Boston, MA: Beacon Press.

Polletta, F. & Jasper, J. M. (2001). Collective identity and social movements. *Annual Review of Sociology*, 27, 1, 283–305.

Reynolds, D. B. (1992). A revolutionary vanguard? Lenin's concept of the party. *Nature, Society, and Thought*, 5, 2, 133–160.

Rosset, P. M. & Martinez-Torres, M. E. (2012). Rural social movements and agroecology: Context, theory, and process. *Ecology and Society*, 17, 3, 17, Online. Retrieved 16 July 2017 from http://dx.doi.org/10.5751/ES-05000-170317.

Routledge, P. (2003). Voices of the dammed: Discursive resistance amidst erasure in the Narmada Valley, India. *Political Geography*, 22, 3, 243–270.

Rucht, D. & Neidhardt, F. (2002). Towards a 'movement society'? On the possibilities of institutionalizing social movements'. *Social Movement Studies*, 1, 1, 7–30.

Snow, D. A. & Benford, R. D. (1988). Ideology, frame resonance, and participant mobilization. *International Social Movement Research*, 1, 1, 197–217.

Snow, D. A. & Benford, R. D. (1992). Master frames and cycles of protest. In A. D. Morris & C. M. Mueller (Eds). *Frontiers in social movement theory* (pp. 133–155). New Haven, CT: Yale University Press.

Snow, D. A. & Oliver, P. E. (1995). Social movements and collective behavior: Social psychological dimensions and considerations. In K. S. Cook, G. A. Fine, & J. S. House (Eds). *Sociological Perspectives on Social Psychology* (pp. 571–599). Boston, MA: Allyn and Bacon.

Snow, D. A, Soule, S. A., & Kriesi, H. (Eds) (2007). *The Blackwell Companion to Social Movements*, Oxford: Blackwell Publishing.

Snow, D. A., Rochford Jr, E. B., Worden, S. K., & Benford, R. D. (1986). Frame alignment processes, micromobilization, and movement participation. *American Sociological Review*, 51, 4, 464–481.

Snow, D. A., Vliegenthart, R., & Corrigall-Brown, C. (2007). Framing the French riots: A comparative study of frame variation. *Social Forces*, 86, 2, 385–415.

Staggenborg, S. (1995). Can feminist organizations be effective? In M. M. Ferree & P. Y. Martin (Eds) *Feminist Organizations: Harvest of the New Women's Movement* (pp. 339–355). Philadelphia, PA: Temple University Press.

Stedile, J. (2002). Landless battalions. *New Left Review*, 15 (second series), 77–104.

Tarrow, S. (1996a). States and opportunities: The political structuring of social movements. In D. McAdam, J. D. McCarthy, & M. N. Zald (Eds) *Comparative Perspectives on Social Movements: Political Opportunities, Mobilizing Structures, and Cultural Framings* (pp. 41–61). Cambridge: Cambridge University Press.

Tarrow, S. (1996b). Social movements in contentious politics: A review article. *American Political Science Review*, 90, 4, 874–883.

Tarrow, S. (2005). *The New Transnational Activism*. New York: Cambridge University Press.

Taylor, D. E. (2000). The rise of the environmental justice paradigm: Injustice framing and the social construction of environmental discourses. *American Behavioral Scientist*, 43, 4, 508–580.

Tilly, C. (1978). *From Mobilization to Revolution*. Reading, MA: Addison-Wesley.

Tilly, C. (1998). Contentious conversation. *Social Research*, 65, 3, 491–510.

Tilly, C. (2004). *Social Movements, 1768–2004*. Boulder, CO: Paradigm Publishers.

Tilly, C. (2005). Introduction to Part II: Invention, diffusion, and transformation of the social movement repertoire. *European Review of History: Revue européenne d'histoire*, 12, 2, 307–320.

Tilly, C., Tilly, L. A., & Tilly, R. H. (1975). *The Rebellious Century: 1830–1930*. Cambridge, MA: Harvard University Press.

Touraine, A. (1974). *The Post-Industrial Society*. London: Wildwood House.

Touraine, A. (2002). The importance of social movements. *Social Movement Studies*, 1, 1, 89–95.

Turner, R. H. & Killian, L. M. (1957). *Collective Behaviour*. Oxford: Prentice-Hall.

Van Der Heijden, H. A. (2006). Globalization, environmental movements, and international political opportunity structures. *Organization & Environment*, 19, 1, 28–45.

Vergara-Camus, L. (2013). Rural social movements in Latin America: In the eye of the storm. *Journal of Agrarian Change*, 13, 4, 590–606.

Wagner, A., Polak, P., & Świątkiewicz-Mośny, M. (2019). Who defines – who decides? Theorising the epistemic communities, communities of practice and interest groups in the healthcare field: a discursive approach. *Social Theory & Health*, 17, 2, 192–212.

Weible, C. M. & Sabatier, P. A. (2006). A guide to the advocacy coalition framework. *Handbook of Public Policy Analysis. Theory, Politics, and Methods* (pp. 123–136). Boca Raton, FL: CRC Press.

Wolford, W. (2010). Participatory democracy by default: Land reform, social movements and the state in Brazil. *The Journal of Peasant Studies*, 37, 1, 91–109.

Zavestoski, S., Agnello, K., Mignano, F., & Darroch, F. (2004). Issue framing and citizen apathy toward local environmental contamination. *Sociological Forum* 19, 2, 255–283.

Zhouri, A. (2004). Global-local Amazon politics conflicting paradigms in the rainforest campaign. *Theory, Culture & Society* 21, 2, 69–89.

2   Peasant collective action against disembedding land

The case of Niassa Province, Mozambique

Kajsa Johansson

Introduction

Since independence in 1975, peasant collective actions have been rare occurrences in Mozambique. While there has been discontent it has mostly been expressed in everyday resistance, as individual and family-based actions against private sector firms and the government (Isaacman, 1975, 1980). There have been some collective efforts to influence policy making such as the making of the revised land law which was probably one of the most participatory law-making processes in Mozambique since independence (Johansson & Nhampossa, 2012). Yet, in recent years there has been a change. One of the areas where change can be observed is the province of Niassa in northern Mozambique. In this chapter, the emergence of collective action by peasants in Niassa is examined. The aim is not only to document the emergence of the collective action, but also to theorize why it emerged in this place at this specific moment. The chapter is based on ethnographically inspired fieldwork carried out between 2014 and 2016. The empirical material of the chapter was gathered during two main field visits in Niassa province and Maputo in 2015 and 2016. Pilot fieldwork was carried out in 2014 in Nampula province and in Maputo. Interviews and life stories were conducted with peasants, peasants' organizations, community leaders, government officials and private sector representatives, including forestry investors. Participatory observations were carried out during peasants' unions' assemblies and meetings and public consultations regarding foreign investments. Before initiating the research, I was working as a development practitioner in Niassa and some examples from that point in time also appear in the text.

Movement dynamics

Emergence

Niassa is Mozambique's largest and least densely populated province with approximately eight people per square kilometre. The province has fertile soils, good rainfall and a suitable temperature for agricultural production.

The vast majority of the population are peasants. On average, each peasant family cultivates 1.5 hectares, mainly for household consumption, through slash-and-burn rotational agriculture. Despite the favourable conditions, productivity is low due to a lack of improved seeds, rain dependency, lack of irrigation schemes, poor extension services, and little use of draught animals. The main, and in most cases only, tool used in the fields (*machambas*) is a hoe. A majority of peasant families are engaged in additional income-generating activities, such as brick producing or bicycle repairing. A small portion of peasant families are involved in contract farming of cotton, tobacco, sesame or other crops. Poverty indicators show stagnant levels of poverty at slightly over 50 per cent, but also a deepening poverty gap.

Agriculture in general, and small-scale agriculture specifically, has not been a priority in government or donor policies and investments. The limited efforts made to strengthen the agricultural sector have mainly focused on large-scale investments, managed by the state or private sector, or have been investments at the ministerial level with limited effects at the local level. Development of small-scale agriculture has been left to the private sector. However, the penetration of private sector actors in Niassa has been weak and sporadic, due to unfavourable and non-profitable conditions with bad roads and communication, scattered populations and limited quantities of surplus produced. Although there have been continuous expressions of discontent among peasants with regard to market conditions and lack of government attention to their main concerns, there was no broad organization to collectively pursue peasants' interest.

Yet in the last decade, this has begun to change. From 2007, peasants in Niassa have been increasingly engaged in collective action, across villages and districts. The emergence coincides with the introduction of large-scale tree plantations. In 2005, the Mozambican government together with international donors, such as Sweden, started to promote large-scale investments in tree plantations in Niassa, mostly exotic species like pines and eucalyptus, to promote development. It is difficult to get hold of exact figures regarding the plantations, but when investments were initiated, Niassa was described as the possible future site for the largest exotic tree plantation in the world. This vision has since been revised. The area to be planted in the province is estimated to be about 480,000 hectares (Åkesson, Calengo, & Tanner, 2009) but the area planted so far is only around 32,000 hectares. The funds invested in the plantations in Niassa originate from Sweden, Norway, Holland, South Africa and the USA, often in complex ownership structures and with lack of transparent and accessible governance structures (Bleyer, Kniivilä, Horne, Sitõe, & Falcão, 2016; Church of Sweden, 2014; Johansson, et al., 2012; PEM Consult, 2011).

The content of the emerging conflicts between communities and companies fall into three main categories. First, there were material issues. Peasants expressed a concern that livelihoods were threatened by the allocation of land to forestry companies. Plantations were established close to peasants' fields, villages and roads, and caused reduced access to animal grazing, collection of firewood and construction material, medicinal plants and other natural resources that peasants depend upon. The second category regards the legal aspects of the companies' actions. For example, the objection was raised that community consultations were not carried out in accordance with the Mozambican legal framework, overlooking the participation of the concerned communities. There were also examples of companies not respecting agreements made with communities in terms of boundaries. Third, there were different cultural understandings of companies and peasants in terms of notions of time, space, and work, as these related to the cultural and social values of land.

At first, actions against the companies appeared as isolated critical voices in different villages. The critical peasants initially thought that the problems they were protesting were confined to their village. Nor was everyone critical in the concerned villages. Those peasant families who received employment opportunities at the companies were of a different opinion. There was also a segment of the population that remained indifferent to the plantations. At that point in time, these actions were not different than previous expressions of individual discontent.

The leadership in the villages, both the traditional leaders and those representing the central government, were in general supportive of the plantations. In the villages where the traditional leaders (*régulos*) were seen as being on the side of the companies, this further worsened the situation. The *régulo* is a key institution in the village, and his or her legitimacy rests on being from and of the community. The *régulo* also has legal obligations to represent and defend the interest of their community according to the land law (see upcoming section on the law). Peasant protests were directed towards the companies, the *régulos*, as well as towards local government representatives.

In 2007, the complaints started to be gathered by a local network of environmental activists called ROADS. At that time, it was difficult to raise concerns with provincial authorities and even more difficult with the national authorities. Those who did raise concerns were said to be against the government, part of the political opposition or against development. At that time, there were local peasant unions in some districts of the province but no umbrella organization representing the peasants at the provincial level. With increasing pressure from the companies, an effort was made to increase mobilization and organization among peasants. The steps began with associations at village level getting together to form district unions. At the end of 2008, a provincial peasants union, União Provincial dos Camponeses de Niassa (UPCN) (Peasants' Union of Niassa Province), was founded and had recruited around 23,800 registered members by 2016. UPCN represents the national peasants' union, UNAC (União Nacional de Camponeses).

Peasant involvement in member-based organizations played a crucial role in launching general criticism against the way investments were carried out, as well as specific criticism of the practice of specific companies. Through the peasants' union, as well as through the efforts of organizations such as ROADS, it became clear that what was first seen as isolated problems in some villages and with some investors, was in fact a pattern in all districts in the province where the plantation investments were taking place.

Actions

What, then, was happening in Niassa? Starting with the most spectacular, there were cases of material destruction. From 2008, peasants in several communities began to destroy plantations by uprooting plants and putting fire to seedlings. Companies suggested that the fires would hurt the communities themselves but slowly came to accept that the fires were an expression of community frustration that would not go away with simple information campaigns. In some places, communities organized road blockades to stop the companies' vehicles from arriving at plantation sites.

While the exact order and magnitude of these actions is difficult to reconstruct, due to scant documentation, we can outline a few examples that give a sense of the type of events that unfolded. In February 2011, for instance, peasants in Ngongoti and Mbandezi, Lago district, expressed their discontent with the representatives of Chikweti Forests of Niassa by threatening them with machetes. The company received a letter from the provincial administration telling them to stop planting in that specific area. In April 2011, peasants in Licole and Lipende, Sanga district, uprooted 60,000 pines in an area of 12,000 hectares that had been planted in the 2009/10 season. In June the same year, the conflict was further aggravated and some infrastructure of Chikweti was burned down (PEM Consult, 2011). On several occasions, peasants were imprisoned for their actions, including in the community of Licole in 2011.

A second category of actions – less conspicuous but no less important – included activities such as documentation, training of local communities, and dialogue between different stakeholders. Collecting stories from distant communities was one of the functions of civil society organizations. These stories were made available to national organizations contributing to their legitimate involvement in the national policy dialogue on land-based investments. The organizations trained local communities in the land law, which spurred further mobilization. The organizations were accused, by government and companies, of seeking to mobilize communities against the investments, simply by informing people about the existing legal framework. ROADS, ORAM (Organizacão Rural de Ajuda Mutua – Rural Mutual Aid Organization), as well as UPCN, strengthened their own capacity, and thus credibility, through collaboration with academic researchers, consultants, and international organizations, who shared their knowledge on how to make rudimentary surveys as well as compile and analyse information. Peasants, both at district and provincial level, made attempts to present their critique to companies and government, but were rarely received.

Until 2010, the debate over the plantations in Niassa had mainly taken place within the province, but at the end of 2010 Chikweti Forests of Niassa appeared in the Swedish newspaper *Dagens Nyheter*, claiming they were not involved in any land conflict in Mozambique. After that, the local and national organizations, increasingly started to use international media to get attention to their case as, apparently, there was no solution in sight through local debates. Articles were published in, among others, Swedish, Norwegian, and Dutch newspapers. Some international organizations wrote reports and articles and spread information through their networks.

With the international attention, it became possible for the local movements to speak directly to politicians from the investor home countries, including the Norwegian Minister of Development and Environment (in 2010) and the Swedish Minister for Development Cooperation (in 2011), both during visits to Niassa. The conflicts between forestry companies and peasants declined once the companies stopped expanding and became less visible in the province. However, collective action to defend land rights continued. A more recent example is the resistance against ProSavana; a planned joint large-scale investment between the Mozambican, Japanese, and Brazilian governments and the private sector. The initial plan included plantations of, for instance, soya beans on areas of six million hectares in northern Mozambique.

A third category of actions was the increased mobilization and organization of peasants, both in unions and in informal constellations. This has taken place continuously through trainings on internal democracy, accountability and participation within member-based organizations. Actions to increase organization have also included development of more short-term development projects of relevance to peasant families, including chicken and goat breeding and adult education.

Outcomes

There is, as always, a need for caution when talking about outcomes as results of specific actions. Instead of claiming attribution, i.e. that the actions discussed lead directly to certain outcomes, it is more reasonable to discuss contribution; that the actions contributed to certain developments, jointly with other actions and circumstances. In the latter, global financial crises as well as changes in development policy, political leadership at provincial and national level in Mozambique, ownership and leadership of companies, and donor policy, could be considered as other contributing factors.

One category of outcomes regards changed relations in the province. When civil society and local communities first raised concerns regarding investments, they were denied spaces to talk and were labelled as oppositional and against development. However, over time this changed. Civil society was increasingly granted a seat at the table with government and companies. Views that at first only appeared in civil society reports or in reports of independent consultants, started to appear also in government official accounts. According to civil society and local communities, the threat of burning and destroying plantations and machinery of the companies, forced the companies to come to the table and listen.

At the time of the first reports of rights violations, local communities and civil society activists were isolated in their struggle against traditional leaders who cooperated with district and provincial government officials. However, this changed over time. Traditional leaders realized that they had made a bad trade, giving away the land, many times for comparably small personal benefits such as employment, against the will of their communities. The leaders faced a crisis in legitimacy and some even resigned. Johansson and Persson (forthcoming) provide a more in-depth analysis of the legitimacy of traditional leaders. Traditional leaders increasingly appeared alongside civil society criticizing the investments. Similar changes can also be seen within government structures.

It was difficult to find district administrators openly criticizing the early investments but this became rather common later on. Even at provincial level, criticism came to be expressed against what was seen as policy decided in Maputo without the participation of the province concerned. As one example, at a seminar on land related conflicts in Lichinga in 2011, the district administrator from Lago suggested that the ministry of foreign affairs in Maputo should organize courses for foreigners before they are allowed to come to the districts so that they can learn about Mozambique and its history, society, and culture, including respect for the local communities.

There are several examples where critique of the plantations similar to the ones presented by the peasants have been presented by persons who in earlier years were either proponents of the plantations, or quiet. At the ProSavana provincial consultation in Lichinga in 2015, the director of Green Resources stood up, expressed that he was nervous to speak up but that he had to, as a citizen, as an implementer of agriculture projects, and third, as a producer. He criticized the government for not taking into account the perspectives of the peasant unions and argued that the government has to consider the sovereignty of the communities and the risk of long-term promises that might not be met.

The official government position was still that large-scale private investment would 'bring' development. During a meeting with a middle level manager at the provincial agriculture authority in 2015, the manager was at first praising the large-scale projects. When later on he was asked what he believed they will bring in terms of development for the majority, his reply was that he has no idea, that the projects were conceived in Maputo and simply arrive in the province.

A second category of outcomes concerns the companies. Several of them went through processes of change in terms of management and investment plans. By 2016, only two out of six companies were left in the province. These were Florestas do Niassa and Green Resources. One example is GSFF (Global Solidarity Forest Fund), the investment fund behind the company Chikweti Forests of Niassa. Critique of the company's activities in Niassa started in 2006–2007. In 2009, the problems were deemed acute and the board asked its president to step down. In 2010, a more comprehensive process of change started with re-structuring of the board and management, bookkeeping was revised, a social fund was created, and land titles were revisited. The company started to accept consultations with communities and civil society organizations (Church of Sweden, 2014). Chikweti was acquired by Green Resources in 2014.

Some companies closed down due to difficulties faced, while others merged. The difficulties faced were likely diverse but it is likely that the unexpected resistance from local communities was a contributing factor. Studies of the global rush for land around 2008–2009 showed that countries with weak institutions were preferred by investors taking advantage of weak enforcement of legal frameworks (Alden Wily, 2011; Anseeuw, Boche, Breu, Giger, Lay, Messerli, & Nolte, 2012). In Niassa, new land is no longer granted to companies, causing a 'rupture' in planned plantations with pieces in the middle missing. Existing plantations remained but there has been little or no planting of new forests going on in Niassa, even though planted area is very far from reaching the areas initially planned. Companies have heavily reduced their workforce, from some 9,000 staff to a couple of hundred.

Communities of Niassa now have new (unwanted) neighbours – thousands of pine and eucalyptus trees – but the communities have managed to halt more new ones coming. However, government and private initiatives that would benefit peasants and develop peasant agriculture and livelihoods in general are still largely absent. Hence, little has changed in terms of isolation and abandonment.[1] The level of awareness and mobilization of peasants has changed. The reactions against the aforementioned ProSavana provide one example. Joseph Hanlon (2016) suggests that the 'No to ProSavana' campaign has been one of Mozambique's most successful and demonstrates that an alliance of local groups and international NGOs can change policy. Other scholars also state that the popular resistance in defence for land-rights has been, and will continue to be, key in defining investments in the country (Wise, 2016). The structures among the peasants, that emerged to respond to the large-scale forest plantations, have been able to give an effective response to ProSavana, and hold the project back.

Unlike previous expressions of discontent that were limited in scope and time, the large-scale plantations have caused something to emerge that has survived its initiating cause. This is unusual in Mozambique. A new form of collective action has emerged in Mozambique as a response to plantation developments. Several reports and articles have described many cases of peasants' protests against natural resource investments in Mozambique over the past decade, providing additional evidence on the appearance of a new level of collective action: see for example Åkesson et al. (2009), Sitoe (2009), Chilundo & Cau (2010), Mole, Monteiro, & Quan (n.d.), PEM Consult (2011), Bleyer et al. (2015), Church of Sweden (2014), FIAN (2012), Johansson et al. (2012), Overbeek (2010), and LexTerra (2016). There are, however, few contributions on how this apparent change can be understood theoretically. The remainder of this chapter elaborates on possible explanations based on some key concepts from peasant studies as well as from Karl Polanyi's analysis of commodification.

Some theoretical notions: land, peasant logic, and disembedding

From our account so far, it would seem reasonable to argue that land issues arising from large-scale capitalist penetration of Mozambican agriculture were crucial to the emergence of collective action. This is indeed suggested by the content of the conflicts, as they focussed on how plantations have infringed upon material, legal and cultural aspects of peasant life. This is where any analysis must start, if we are to understand the emergence, actions and outcomes of collective action during the last decade in Mozambique.

Henry Bernstein has argued that, 'the widespread conversion of land into property – into a commodity – is one of the defining characteristics of capitalism' (2010, p. 23). While this assertion is in line with our earlier remarks, the Mozambican case offers some subtleties. As we will see in more detail in the next section, Mozambican law does not allow private ownership of land, thus precluding land commodification in a strict and narrow sense. Within this legal framework, however, provisions have been made to accommodate capitalist penetration of agriculture and the rural sector through the granting of plantation concessions, thereby installing a property regime that is equally effective in reserving areas for capitalist production and in barring peasants from utilizing the land.

In order to capture such subtleties, it is useful to revisit Karl Polanyi's analysis of the great transformation and examine the conceptual tools it yielded. His ambition was to capture the *nascent* conversion of land into property in the English nineteenth century and to treat it not as a one-off event but rather as generating on-going movements and counter movements. Polanyi defined commodities as 'objects produced for sale on the market' (2012, p. 75). Based on this definition, Polanyi identified labour, land, and money as fictitious commodities. They cannot be commodities, he argued, because they are not produced for a market; they are actually not produced at all. Polanyi writes:

> labor and land are no other than the human beings themselves of which every society consists and the natural surroundings in which it exists. To include them in the market mechanism means to subordinate the substance of society itself to the laws of the market. [...] land is only another name for nature, which is not produced by man [...] The commodity description of labor, land, and money is entirely fictitious.
>
> (2012, pp. 75–76)

However, according to Polanyi, this fiction became the organizing principle of the market society, with devastating effects:

> Nature would be reduced to its elements, neighborhoods and landscapes defiled, rivers polluted, military safety jeopardized, the power to produce food and raw materials destroyed. [...] Undoubtedly, labor, land, and money markets are essential to a market economy. But no society could stand the effects of such a system of crude fictions even for the shortest stretch of time unless its human and natural substance as well as its business organization was protected against the ravages of this satanic mill.
>
> (Polanyi, 2012, pp. 76–77)

Polanyi suggested that history can be understood in terms of transformations of embedding, disembedding, and re-embedding of the economy in society, based upon compromises between movements and counter movements. His example treats the transformation of traditional English society from one in which economic relations were embedded in and constrained by social relations, to a 'market society' where economic functions are disembedded from social institutions and begin to determine social relations. In this transformation Polanyi questioned the assumptions that the market was a uniquely natural form of economic organization of society; that human economic behaviour is inevitably motivated by the goal of maximizing profit; and the idea of self-regulating markets facilitated by the non-intervention of the state.

Due to the insecurity experienced by people as the economy becomes increasingly disembedded, there are reactions, or counter movements claiming the necessity to re-embed the economy in society. Civil society demands that the state protect society from the effects of the market; that political decisions should curb the rampage of the market economy and soften its sharp edges. Applying Polanyi's analysis, Mozambique appears to be in the transition from a traditional to a market society with plantations as the primary economic force disrupting peasant social relations.

Scholars of peasant studies offer a similar insight. They argue that when the logic of the peasant's moral economy is threatened, peasants react. Bernstein suggests that a peasant is a 'household farmer organized for simple reproduction, notably to supply its own food ('subsistence'). Often added to this basic definition are presumed qualities such as the solidarity, reciprocity and egalitarianism of the village and commitment to the values of a way of life based on household, community, kin and locale' (2010, p. 3).

In regards to the peasant logic and land in the case of Mozambique, it is important to underline that the point is not to idealize or claim that traditional leadership and division of land is equalitarian, Jossias (2016) shows that it is not. But the peasant logic does represent another way of seeing things.

Analysis

The Mozambican Land Law, its enforcement, proponents, and critics

The Mozambican Constitution states that all land, water and natural resources belong to the state. Mozambique has had three land laws. The first one was the discriminatory colonial indigenous law, which underwent some changes in 1960 to formally recognize the right of Mozambican 'natives' to own land. The second land law was approved in 1979, in the early years of Independence. When this law was presented in the parliament, the minister of justice Teodato Hunguana claimed that 'overtaking of the land by the Mozambican people was not the result of a land reform, but the culmination of the revolutionary National Liberation Struggle' (quoted in Mosca, 2010, p. 203).

Although the Constitution protected the rights of land users in general, according to the 1979 Land Law, acquiring a title which would formally secure the user's rights was out of reach for the majority of the rural population due to high cost and bureaucracy. In the 1990s, the land law was revised in a process where civil society organizations, including UNAC and ORAM, played an important role. This resulted in the revision of the 1979 Land Law (Johansson et al., 2012).

The approved current law (Law n° 19/97 of 1 October) is considered progressive in its recognition of various forms of rights and in maintaining the principle that land is not a commodity that can be sold. Only the right to use and benefit from, and not to own, land can be acquired. The law recognizes communities' right to hold land titles, i.e. the right to use, control, and inherit land. The law further admits the right to land through long-term occupancy (more than ten years) based on oral testimony of community members, considered to be as strong as written proof of ownership. In case of non-community members seeking land use rights, they are required to consult and negotiate access to land with the local communities (Law n° 19/97 of 1 October; Åkesson et al., 2009; Nhampossa, 2011).

Some of the critics of the current land law argue that it would be favourable for all if land was privatized, pointing to the many conflicts between investors and communities that the law seems unable to avoid or to solve. One of the proponents of this position is a former senior government official, presently an active scholar in agriculture and land issues, claiming that the peasant movement had too much to say in the elaboration of the current land law, and that the result was a law that is not adapted to the actual context of Mozambique:

> everybody says that it is good but no one is following it. If we have a law that is impossible to implement, how can it be good? I would say that it is the worst law and that it doesn't solve anything. The problem is that the law is based upon the idea of a socialist state but it is applied in a world of total capitalism, of savage capitalism. There is a disconnection between the reality and the foundations for the law.
> (Interview, Carilho, 31 March 2014)

However, it could also be argued that this 'mismatch' or anachronism of the law is its strength, as expressed by a former coordinator of UNAC, Diamantino Nhampossa:

> The law can be seen as a platform for struggle. The law was processed during a time when we had already embarked on the road to market economy. But the pride of the history is still present in the law.
> (Interview, Nhampossa, 30 March 2014)

There is yet another angle, presented by Mosca (2010) and Jossias (2016). They claim that there is, in practice, no contradiction between the law and the interests of the IMF, the World Bank, the national élite, and external capital because the law presents, in practice, no limit to usurp vast areas of land. Hence, the present situation allows for a de facto privatization.

Jossias (2016) argues that one of the main principles of the law is to ensure that the state and governmental bodies in Mozambique should have little or no interventionist role in the allocation of land. Land allocation is delegated to the representatives of the traditional leadership referred to in the law as 'local communities', which was a term introduced by the land law but which does not exist in the administrative divisions of Mozambique. Hence, the civil rights of the state become reserved for the urban, while the rural is left with the customary norms and practices, in a 'bifurcated state' (Jossias, 2016, p. 63).

According to the peasants, the state is inactive or absent and fails to enforce the parts of law that defends the rights of local communities, while, active on the side of the investors. Many of the traditional leaders were at first also on the investors' side and were accused of having been corrupted by the companies' offers for short-term personal economic gain, such as employment opportunities for them and their families (Åkesson et al., 2009; Church of Sweden, 2014). This led to a disruption in the local governance structure that, in the absence of the state, is key for everyday life in the local community. The communities were left with nobody to represent them.

Land as livelihood

The dispossession of land, as it has occurred in Niassa, separates peasants from their means of subsistence and neglects the many ways peasant communities depend on land and natural resources. The promise to trade land for jobs takes advantage of peasants' precarious living conditions and increases their dependency on cash income. It is crucial to acknowledge that peasant families do not just need and use the land for farming on their *machambas*; they depend on vast areas of land to cover the need for several natural resources that are key to their survival, including building material, grazing lands, medicinal plants, water, hunting and beekeeping. The commons, or community lands, are central to the livelihoods of all families but most important for the poorest. When pressure on land increases, this is usually the land most difficult to protect (Alden Wily, 2011; Åkesson et al., 2009; Johansson & Åkesson, 2014). The practice of slash-and-burn rotational agriculture, where land is left in fallow for years or even decades but is still considered to be a part of a family's planting system, further increases the area that each family needs to secure its livelihood. Peasant families' need for land is often underestimated and many times reduced to the area of the *machamba*.

When the plantations were established in Niassa, several investors promised that the community by giving up land would get jobs. In the discourse of investors, governments and donors, there was an assumption of an apparent automatic translation of job opportunities into livelihood improvements. It is difficult to find any proof of this but increased precarity and insecurity in terms of incomes have been highlighted. Several sources (Bleyer et al., 2016; Åkesson et al., 2009; PEM Consult, 2011, and ORGUT, 2006) provide evidence that there has been a lack of livelihood improvements resulting from large-scale investments.

The message from the companies on land and jobs was perceived by many citizens as an exchange; land was given and jobs would be received. But the jobs were not permanent; many were seasonal, coinciding with the most intense farming season in family agriculture, and even where jobs were full-time, they only lasted for a couple of years, during the land preparation and planting phase. The land, on the other hand, was given away to the plantations for 49 or 99 years. When rural citizens look back, land appears as a far more faithful partner than the promise of possible short-term wealth provided by a company.

It is in this regard important to see the plantation investments from a historical perspective. Niassa is characterized by internal and external isolation. Even though people probably knew that the investments would not have a long-term positive impact on the lives of the many, it was at least *something*. A *régulo* in Sanga district explained that they felt that nothing happened after the Peace Accords in 1992 until the forestry investments arrived. These represented a possibility for change.

> And now it has been planted. It is planted. If they are not managing, it is their fault, we have given the land away. And now there is no employment anymore, that was not what they promised in the beginning. There are some seasonal workers. If they [the company] would come back and ask for more land we would not give them. There is no space. We don't want more. If they would grow food yes, but enough with the pine trees. [...] The area that we gave to the companies is big [...] The deal was that the project would bring development.
>
> (*Régulo* Calanje, 9 May 2015)

Similar accounts are given by other inhabitants in the areas of the plantations. Another *régulo* in the same district says that nothing has become what they expected:

> this is employment that brings hunger [...] what we say now is that we want food. But now we are here with all these pine trees. Now we are saying no. You can ask also in other places. These logs are in the whole province. But do you really think that all these people accepted this?
>
> (*Régulo* Chipanga, 9 May 2015)

Peasants state that they have been feeding themselves for centuries; no one else has ever fed them – making the land, and secure access to it, the foundation for survival. The increase of large-scale land acquisitions threatens this modus vivendi, not just through the decreased access to land for peasant agriculture, but also through its short-term perspective of providing jobs and cash-income. Peasant agriculture functions as a sponge or reserve of labour, as pointed out by Castel-Branco (2014) for the prioritized sectors of the economy. But the statements of the peasants show that peasant agriculture is more than this. Changing land access restructures all aspects of peasant life.

How the loss of land access and compensation for peasants is conceptualized by the companies can be found in the companies' discussion of social responsibility. Assets that are not comparable were compared: Peasants' lost their land but where 'given' schools, maternity wards, and ambulance bicycles. The Mozambican intellectual Brazão Mazula describes this perspective:

> It is like it [the land] was just about access to services, like a hospital. But even if they have that, they have been distanced from their fields of production. And the big companies don't absorb the peasants as workforce. And then they compensate by providing water and health but they overlook the natural right of man; to work. You will find people sitting on the veranda of a newly built house in Tete, that is indeed a lot better than the *pau-a-pic* [traditionally made house of bamboo sticks, small stones and clay] that they had before, but they have no job and no land.
>
> (Interview, 9 June 2015)

Land as identity, belonging and history

Interviews with peasants provide insights into the deeper and underlying conflicts between their previous modus vivendi and the one created by the investors. The logic of the investments is that although people might be hungry, the pine and eucalyptus trees have a superior value. When leaving a peasant focus group meeting in Bagarila, Sanga an old man approached me and said: 'I prefer to live in misfortune than in an invaded community. I don't accept any longer this idea of the pine. I simply don't accept it' (Focus group interview, 10 May 2015).

Companies claim to be planting on marginal lands but to peasants there is no marginal land, the assembly of these two words appears a contradiction in itself. Land might not be planted, but it is used, and even if it is not, calling it marginal is to marginalize the people who live, and lived, from it. During a confrontation between the management of the forestry company Chikweti in 2010, Paulino Imede, the coordinator of the Lichinga peasant union, responded as follows when the company claimed they only used wasted and unproductive land for their plantations:

> I don't accept that you call the land you are planting on unproductive and useless. Our families have survived on these lands for centuries, it has given us everything we have and everything we are.

The statement exemplifies that land is not only related to, but equated with, the ancestry and the being of the peasants. The narratives of the land are the history of the country, the foundation of the territory and the identity of those residing in it.

There is a general expression among informants, they identify themselves with pride as 'children of the land' (in Portuguese *filho do camponês*).[2] When peasants define those among the élite who are on the side of the peasants, they are explained to be so since they are also *filho do camponês*.

Yet another dimension of land, where the peasant logic is at odds with the market society, is the question of land as separable into independent pieces. To peasants, the land is one. Hence, when land rights of any peasant community are threatened, it affects everyone, not only in that community but in others because they share history and identity. In the communities, the land is divided between peasant families but the division is something that is embedded in the social structures of that community. The land belongs to the collective. During a peasant and fisher-folk focus group in Meluluka, Lago district, it was explained how land is distributed, delimitation is done, and how they know which land belongs to which family in the community. There is no investor operating. They explained by drawing maps of the *machambas* of António, João, and Samuel in the sand. They claim that there is a total clarity of the maps and land allocations. I asked where these maps are and they responded: '*Estão nas cabeças dos machambenses*' (They are in heads of the farmers).

When this undivided land is threatened by division, it represents to peasants, an act of violence. The violence is expressed when the entrance of the large-scale investors is described, often using concepts like 'invasion' and, like the president of the provincial peasant union, stating that land deals are being made 'using the name of Mozambique pledging our land using a knife' (Interview, 24 May 2015).

Most field interviews focused on the period in time since independence in 1975, and informants of the older generation related their life stories to the struggle for the land that took place in the struggle for independence. Peasants define themselves as part of, or rather owners of, the liberation struggle; as fighters, providers of food or other necessary support. The interviews show how the struggle and the liberation defined the territory and hence the land, representing not only where people live and make their livelihood, but also who they are, what they do and what they struggled for. The peasantry of Niassa was deeply involved in the struggle for independence and there is a significant identification with Frelimo as the movement liberating the land. The peasants are proud and express ownership over it, but at the same time, they express despair and disappointment due to the lack of attention and benefit received since independence.

During the liberation struggle, the land enjoyed a key position in the rhetoric of what should be liberated; the man and the land, not as two but as inseparable. From the accounts of the peasants, we learn that for them, these were not merely slogans but something that was and is deeply attached to what they actually fought for. And when independence was gained that is what they expected to have; a liberated land in all its dimensions. The founder of UNAC, Ismael Ossemane, refers to the period at the end of the 1980s when the market economy was introduced in Mozambique:

But with the market economy, conflicts started to arise that had been unknown to us before and the issue was raised about the need to have another land law. To gain a title was a bureaucracy and you needed money. But the peasants had neither money nor the capacity to pursue a case. But even more, to them it was impossible to even imagine that someone could take what was closest to them, their land, through a piece of paper.

(Interview, 9 May 2014)

The right to and relation with land can be equated to the peasants' (national) identification; the *moçambicanidade* and belonging. As stated by Mazula:

for the peasant, the land is like their identity card, their BI [Bilhete de Identidade]. The market comes with a piece of paper, but the peasant will say that this is the land of my ancestors. The government is trying to reduce the conflict through giving DUATs [land titles] [...] Out of the 75 per cent of our population that are peasants, how many have a real BI? But what says that he is a Mozambican is the land.

(Interview, 9 June 2015)

In addition to showing the violation represented by taking land for a plantation, the quote also relates to the role of Government; selling and consuming the land of its people, or, if land and people are seen as insep-arable, selling its people. In Cuamba, feeling that they are being ques-tioned as land owners led a group from the district peasant union to give the following account:

In politics they say that the land belongs to nobody. So we say that it is ours. But in the big meetings they say that the land belongs to the People, and we are the People. ... There is no clear information between the government and its People, but in politics they say that the Government is the people. Maybe if Eduardo Mondlane [Frelimo's first president] wouldn't have died, with the project that he had things could be different [...] The People want change and the government says things can't change, they have to remain since I am the one governing and when things like that happen, a revolution could emerge. That is what we are waiting for, one of these days.

(Interview, 25 May 2015)

Several interviews, including the one in Cuamba, exemplifies how peasants are aware of and engaged in rights issues and politics, resulting from the increased pressure on the land. Ismael Ossemane, the founder of the national peasants' union, UNAC, gives the following account regarding the land law and awareness of rights:

When we were campaigning for the new land [i.e. the 1997 law], that was before the severe problems that we have now started, the peasants even fell asleep in our meetings. But today it is very difficult to find a meeting about land rights with peasants falling asleep; no one will sleep in the areas where they have some experience with the megaprojects.

(Interview, 9 May 2014)

A counter movement against disembedding

The struggle in defence of land rights is a rare case in Mozambique where the collective interests of the peasants, the citizens of the countryside, have generated collective action. The Mozambican land law prevents turning land into a commodity but at the same time, land is a part of the broader market society. The effect on peasants' lives of the way that land is being dealt with are the same as if land has been commodified. We are able to understand why the dispossession of land, unlike the many other injustices suffered by peasants, can cause collective action and not 'just' everyday individual struggle. The peasants' struggle for land goes beyond a defence of farming practices and concerns their modus vivendi; the logic, of the peasant community. When the cotton or tobacco company fools the peasant with the scale; the day-labour is not paid what was agreed; no one comes to buy your charcoal, or the government does not bring the inputs promised, these will be difficult, but they will not threaten their being as peasants, including their history, belonging and identification.

Peasants' collective action in the case of Niassa can be understood as a countermovement against the disembedding of the economy, rather than strictly the commodification of land. The land deals are not only something foreign to the peasant logic, they actually threaten the possibility to continue life according to that logic, where the map of the land resides in the minds of the *machambenses*. Given civil society's earlier victories in terms of protecting the land from commodification in the land law, the current struggle can be viewed as a struggle against the bifurcate state; it is a call for the state to provide protection to all its citizens from the sharp edges of the market knife.

Three decades ago, James Scott commented on the effects of the Green Revolution on peasants in Kedah, Malaysia. Scott argued that the Green Revolution served to diffuse class conflict because,

it simply removes the poor from the productive process rather than directly exploiting them [...] If the profits of the green revolution had depended on squeezing more from the tenants, rather than dismissing them, or extracting more work for less pay from labourers, the consequences for class conflict would surely have been far more dramatic.

(Scott, 1986, p. 13)

Mozambique is experiencing a similar tension. It is not engaged in a class conflict but rather in a struggle against the disembedding of the economy from peasant society. Erik Olin Wright claims that an underclass 'consists of human beings who are largely expendable from the point of view of the rationality of capitalism' (1994, p. 49). The peasants of Niassa are not an underclass in the sense that they are exploited on their land but they are in fact expendable from the perspective of the large-scale plantations. During the first years, there were thousands of employment opportunities on plantations, but they declined quickly in a couple of years. As Olin Wright explains:

> In nonexploitative economic oppression there is no transfer of the fruits of labor from the oppressed to the oppressor; the-welfare of the oppressor depends simply on the exclusion of the oppressed from access to certain resources, but not on their laboring effort.
> (Olin Wright, 2000, 10f.)

In the exploitative relation, the exploiter depends upon the work of the exploited but '*In the case of nonexploitative oppression, the oppressors would be happy if the oppressed simply disappeared*' (Olin Wright, 2000, p. 11). In terms of power it is worse to not even be needed, than to be exploited. After just a couple of years, the communities became painfully aware, and surprised, that their work force was not needed by the investors once cleaning of land and planting was done. Hence, the rapid transition from being exploited to being expendable is built into the kind of plantation development taking place in Niassa, as well as elsewhere in Mozambique, and in many other countries. The exploitation phase is merely a very brief step towards the much longer-lasting oppression of expulsion.

There is an on-going debate within peasant studies (see, for example, Bernstein, 2016) whether peasants are engaged in class struggle with agribusiness in the form of large-scale land investors. This chapter suggests that peasants' struggle should not primarily be understood as a class struggle, but as a civil society countermovement in a Polanyian sense, against the disembedding of the economy. The disembedding process driven by forestry investments, makes the peasants and their society unnecessary and expendable and this in just a few short years.

Notes

1 In October 2017, a delegation composed of representatives from UNAC, UPCN, the district peasants' union in Ribaue, Nampula province, Justiça Ambiental and Livaningo made a visit to Norway and Sweden, invited by Afrikagrupperna. The aim was to raise awareness on the present effects of the plantations and to dialogue with politicians and investors on the issue. The

issue of hopelessness and uncertainty was recurrent in the statements of the peasants.

2 *Camponês* comes from the word *campo*, which means field, land, or rural area. Hence, *camponês* is someone living in and from the land.

References

Åkesson, G., Calengo, A., & Tanner, C. (2009). It's not a question of doing or not doing it – it is a question of how to do it, Study on Community Land Rights in Niassa Province, Mozambique, SOL No. 6/2009. Uppsala: Institutionen for stad och land, the Swedish University of Agricultural Sciences.

Alden Wily, L. (2011). *The tragedy of public lands: The fate of the commons under global commercial pressure.* ILC Collaborative Research Project on Commercial Pressures on Land. Rome: International Land Coalition.

Anseeuw, W., Boche, M., Breu, T., Giger, M., Lay, J., Messerli, P., & Nolte, K. (2012). *Transnational Land Deals for Agriculture in the Global South. Analytical Report based on the Land Matrix Database.* CDE/CIRAD/GIGA, Bern/Montpellier/Hamburg.

Bernstein, H. (2010). *Class Dynamics of Agrarian Change.* Halifax: Fernwood Publishing.

Bernstein, H. (2016). Agrarian political economy and modern world capitalism: the contributions of food regime analysis. *The Journal of Peasant Studies*, 43(3): 611–647.

Bleyer, M., Kniivilä, M., Horne, P., Sitõe, A., & Falcão, M. P. (2016). Socioeconomic impacts of private land use investments on rural communities: Industrial forest plantations in Niassa, Mozambique. *Land Use Policy*, 51, 281–289.

Castel-Branco, C. N. (2014). Growth, capital accumulation and economic porosity in Mozambique: social losses, private gains. *Review of African Political Economy*, 41, 1, 26–48.

Chilundo, A. G. & Cau, B. M. (2010). *Traditional Forms of Common Property Rights: A Case Study in Southern Mozambique.* Maputo: Universidade de Eduardo Mondlane.

Church of Sweden. (2014). *Assessment visit to Chikweti, Niassa. Report from the joint delegation visit 15–23 October 2013.* Uppsala: Church of Sweden.

FIAN. (2012). *The Human Rights Impacts of Tree Plantations in Niassa Province, Mozambique.* Heidelberg: FIAN.

Governo de Moçambique (1997). *Lei de Terras de 1997 de 1 de Outubro.* Maputo.

Hanlon. J. (2016). What does a successful campaign do after it wins? *Mozambique News Reports & Clippings*, 329. 26 June 2016.

Isaacman, A. F. (1975). The tradition of resistance in Mozambique. *Africa Today*, 22, 3, 37–50.

Isaacman, A. F. (1980). Cotton is the mother of poverty: Peasant resistance to forced cotton production in Mozambique, 1938-1961. *The International Journal of African Historical Studies*, 13, 4, 581–615.

Johansson, K. & Åkesson, G. (2014). Community structures and the governance of community land under pressure: Mozambique 15 years after approval of the land law. Unpublished conference paper.

Johansson, K. & Nhampossa, D. (2012). Land tenure and governance. In K. Gregow, K. Hermele, K. Johansson, D. Nhampossa, & M. Wolpher (Eds) *The Race for*

Land (pp. 25–30). Stockholm: Afrikagrupperna/Forum Syd/Swedish Cooperative Centre.

Johansson, K., Nhampossa, D., & Wolpher, M. (2012). Peasants' voice having an impact. In K. Gregow, K. Hermele, K. Johansson, D. Nhampossa, & M. Wolpher (Eds) *The Race for Land* (pp. 35–39). Stockholm: Afrikagrupperna/Forum Syd/ Swedish Cooperative Centre.

Johansson, K. & Persson, M. (forthcoming). Olika regimer, samma ledarskap – Om det lokala maktfältets logik på landsbygden i Moçambique. *Praktiske Grunde.*

Jossias, E. M. F. (2016). O primeiro a chegar é dono da terra: pertença e posse da terra na região do lago Niassa. Universidade de Lisboa. Unpublished PhD dissertation.

LexTerra. (2016). *The Progress of Forest Plantations on the Farmers Territories in the Nacala Corridor: The case of Green Resources Mozambique.* Maputo: Justiça Ambiental, Livaningo and UNAC.

Mole, P., Monteiro, J., & Quan, J. (n.d.). *Ensuring community land rights in a land investment pressed country – The Community Land initiative (iTC) in Mozambique.* Maputo: iTC.

Mosca, J. (2010). *Políticas Agrárias de (em) Moçambique (1975–2009).* Maputo: Escolar Editora.

Nhampossa, D. (2011). Genesis and role of the peasant movement in Mozambique in K. Helliker & T. Murisa, (Eds) *Land Struggles and Civil Society in Southern Africa* (pp. 191–216). Trenton, NJ: Africa World Press.

Olin Wright, E. (1994). *Interrogating Inequality: Essays on Class Analysis, Socialism, and Marxism.* London: Verso.

Olin Wright, E. (2000). *Class Counts* (student edition). Cambridge: Cambridge University Press.

ORGUT (2006). *Linkages between livelihoods and natural resources.* Mozambique. Stockholm: ORGUT.

Overbeek, W. (2010). *The Expansion of Tree Monocultures in Mozambique: Impacts on Local Peasant Communities in the Province of Niassa. A field report.* Montevideo: World Rainforest Movement.

PEM Consult (2011). *Estudo Sobre a Gestão dos Conflitos de Terra Entre as Comunidades e Investidores nas Plantações Florestais da província de Niassa.* Maputo: PEM Consult.

Polanyi, K. (2012). *Den Stora Omdaningen.* Lund: Arkiv förlag.

Scott, J. C. (1986). Everyday forms of peasant resistance. *The Journal of Peasant Studies,* 13, 2, 5–35.

Sitoe, A. A. (2009). *Governação Florestal em Niassa: o caso de Muembe, Sanga, Lago e Cuamba.* Maputo: ORAM – Rural Association for Mutual Aid.

Wise, T. A. (2016). Land Grab Update Mozambique, Africa Still in the Crosshairs. *Food Tank.* Retrieved 31 October 2016 from https://foodtank.com/news/ 2016/10/land-grab-update-Mozambique-africa-still-in-the-crosshairs/.

Interviews

Bagarila peasants focus group, 10 May 2015, Sanga.

Baptista Amid, João, historian and pedagogic director at Universidade Católica de Moçambique, Niassa. Lichinga, 28 May 2015.

Carilho, João, researcher affiliated to OMR, former deputy minister of agriculture, 31 March 2014, Nautilus, Maputo.
Cuamba district peasants' union, 25 May 2015, UCM, Cuamba.
Ismael Ossemane, founder of UNAC, 9 May 2014, UNAC's office, Maputo.
Mazula, Brazão, doctor in philosophy, previously director of Universidade Eduardo Mondlane Maputo, born in Messumba, Niassa. 9 June 2015, UEM, Maputo.
Middle level manager, Provincial Agriculture Authority, 1 April 2015, DPA, Lichinga.
Nhampossa, Diamantino, We Effect country representative, former UNAC coordinator, 30 March 2014, We Effect, Maputo.
Régulo Calanje, 9 May 2015, Sanga.
Régulo Chipanga, 9 May 2015, Sanga.

3 Negotiating pipeline projects and reterritorializing land through rural resistance in northern Kenya

Charis Enns and Brock Bersaglio

Introduction

In 2012, construction began on the Lamu Port–South Sudan–Ethiopia Transport (LAPSSET) corridor in northern Kenya. The LAPSSET corridor is anchored by a pipeline that will transport crude oil across northern Kenya to a new port facility on the Indian Ocean. In addition to the pipeline and port, the corridor will include a new highway, railway and other transport infrastructure, such as international airports. By some estimates, over 1,500,000 hectares of land will be acquired to bring the project to fruition (Laurance, Sloan, Weng, & Sayer, 2015). The LAPSSET Corridor Development Authority (LCDA) promises that the corridor will catalyze further investments in industrial and agricultural development and the expansion of social and economic services – thereby benefiting rural groups in northern Kenya that have long been neglected by the state.

The development of LAPSSET has been described as part of a larger process of territorial restructuring taking place in northern Kenya (Greiner, 2016). For many decades, the peoples and landscapes of northern Kenya have been treated as marginal or peripheral to the state. However, recent discoveries of oil, geothermal and wind resources have attracted foreign investors to the region. The state has responded to investor interest by beginning to administer and defend territory in the north differently than in the past – reterritorializing land that was claimed during the colonial era but that has since existed within relatively fluid and porous boundaries and borders. In short, a state-led process of territorial restructuring is unfolding in Kenya to facilitate the development of the country's newly discovered natural resource reserves and to ensure that extracted resources can move with ease across the country to global markets.

This chapter considers how rural groups in northern Kenya are being impacted by, responding to, and influencing this process of territorial restructuring for natural resource development. Following a brief discussion of the concept of territorial restructuring, an overview of the study area is provided. Next, LAPSSET and its impacts on rural groups are discussed. Although the LCDA has attached promises of economic and social development to LAPSSET, many rural groups in northern Kenya are concerned about how the corridor is driving changes in land use, tenure, and value, as well as creating new forms of competition over land and resources. Specific strategies being used by rural groups to 'modify, subvert, block and even overturn' LAPSSET related decisions and developments that threaten to restrict access to and control over land and resources are then examined. It is argued that rural groups are engaging in acts of (re)territorialization 'from below' to resist territorial restructuring 'from above' (Borras & Franco, 2013; Brent, 2015; Peluso, 2005). The final section of this chapter raises questions about the outcomes of this approach to rural resistance. Consideration is provided for how efforts to reclaim control over land and resources through acts of reterritorialization can have uneven outcomes in rural society, aggravating historical anxieties, inequalities and points of contention between different types of rural land users. This chapter is informed by data collected during two periods of fieldwork in northcentral Kenya, totalling 11 months, between 2014 and 2017.

Territorial restructuring, natural resource development and rural resistance

In recent years, a renewed and intensified wave of large-scale investments in land and resources has taken place across much of the Global South – a phenomenon commonly referred to as 'the global land rush' (Borras, Hall, Scoones, White, & Wolford, 2011; Deininger, 2011; Margulis, McKeon, & Borras, 2013). The global rush for land and resources has been 'characterized by transnational and domestic corporate investors, governments and local elites taking control over large quantities of land (and its minerals and water) to produce food, feed, biofuel and other industrial commodities for the international or domestic markets' (Margulis et al., 2013, p. 2). This rush for land and resources has created new competition and contestation between various actors, including state entities, domestic and transnational investors, various elites, communities, and non-governmental organizations.

Territorial restructuring is a strategy used to claim access to and control over land and resources that are subject to contestation between actors (Lund, 2011; Hall, 2013; Margulis et al., 2013). 'Territorialization' refers to mechanisms and processes used to *claim* and *restrict* land and resources for certain types of uses by certain types of actors at the expense of others (Corson, 2012; Lund, 2011; Peluso & Lund, 2011; Ribot & Peluso, 2003). Acts of territorialization include mechanisms used to establish control over land and resources or the people who use them, such as: creating and mapping boundaries, allocating or restricting access to land and resources to certain users, or appropriating land and resources to reallocate them from one user to another (Borras et al., 2013; Corson, 2012, p. 705; Vandergeest & Peluso, 1995).

States often deploy acts of territorialization to free land and resources for investment – facilitating investors' access to land and resources by asserting control over a geographical area and the people within it. However, states are not the only actors that can claim land and resources through acts of territorialization. Rural resistance to large-scale investments is often animated by competing notions of territorial rights and sovereignty (Alonso-Fradejas, 2015; Brent, 2015; Corson, 2012; Greiner, 2016; Peluso & Lund, 2011; Rocheleau, 2015). In such instances, rural groups may engage in their own acts of territorialization as a means of negotiating the terms of investment or resisting investment altogether. These acts may include: physically occupying lands, undertaking community land and resource mapping exercises or pursuing formal/legal rights to solidify claims to land. Through such acts of (re)territorialization, rural groups aim to exert authority over how land should be used, by who, and to what end (Alonso-Fradejas, 2015; Brent, 2015; Cavanagh & Benjaminsen, 2015; Rocheleau, 2015).

The analysis that follows examines specific acts of (re)territorialization being used by rural groups in northern Kenya to negotiate or resist components of the LAPSSET corridor. The analysis demonstrates how some rural groups have been able to influence decisions related to LAPSSET – effectively negotiating the state's attempt to restructure territory from above by (re)drawing boundaries and borders from below. However, not all rural groups have been able to achieve the same degree of success in reclaiming land and resources in response to corridor development. It is argued that reasserting territorial control over land and resources from below can have uneven outcomes, as rural groups occupy different spaces and strata in rural society which affords them different levels of influence. In the case of northern Kenya, acts of reterritorialization have helped solidify the territorial claims of some rural groups – such as actors from the conservation sector – with the state now recognizing these groups' claims to land and resources while marginalizing or sidelining the claims of other rural groups to land and resources threatened by LAPSSET.

Contextualizing political reactions to LAPSSET in northern Kenya

Northern Kenya has a long, complicated history of contentious land and resource politics. This history was complicated by the arrival of British colonizers in the late 1800s. Long-standing disputes over territory and resources have shaped the region's political geography, as well as social relationships and inequalities between rural groups. Consideration for the history of land and resource politics in northern Kenya provides important context for understanding the process of territorial restructuring underway today to facilitate LAPSSET as well as the reactions that have emerged from different rural groups in response.

Pastoralism remains a predominant livelihood strategy in northern Kenya. There is a good deal of socio-cultural and economic differentiation among pastoralist groups. However, most pastoralists – including Maasai, Pokot, Samburu, and Turkana, among others – practice similar livelihood activities and use similar land management practices, migrating seasonally along traditional routes to conserve pasture and water (Catley, Lind, & Scoones, 2013; Fratkin, 1997). Because the region is prone to variable rainfall and drought and is covered by sparse vegetation and a desert landscape, much of the population moves seasonally to sustain herds – particularly during drought and other times of scarcity. As a livelihood system, pastoralism is well adapted to the arid and semi-arid conditions of the region (Fratkin, 1997). The socio-cultural practices and knowledges embedded in this livelihood strategy enable people to make use of dryland areas, where the success of other land uses – such as farming – has conventionally been constrained. In addition to being well-suited to the arid environment, pastoralism also makes important contributions to the national and regional economy. Estimates suggest that pastoralism contributes about 13 per cent of Kenya's annual GDP (*New Humanitarian*, 2013). Yet many urban Kenyans and political elites continue to perceive pastoralism as economically unproductive, as well as an obstacle to national development, modernity and progress (Catley et al., 2013; Fratkin, 1997; Salih, Dietz, & Ahmed, 2001).

The marginalization of northern Kenya and of the pastoralists who live there dates back to the colonial era. In the early 1900s, the colonial administration began to allocate public resources disproportionately to highland areas in central and western Kenya perceived as being 'more productive' (Eriksen & Lind, 2009). These areas were also reserved for settlement by white Anglo-European migrants. During this period, many pastoralists were evicted from fertile highlands and pushed south or north so that white settlers could establish commercial ranches for beef production and game reserves for sport hunting (Bersaglio, 2017; German, Unks, & King, 2017; Hughes, 2006; Sundstrom, 2009; Steinhart, 2006). Pastoralists, who traditionally relied on the highlands for dry season grazing, were restricted from moving their herds into the region. The British effectively created an internal border between northern Kenya and the 'white highlands' by building frontier posts to confine pastoralists to *Afrique inutile* while securing *Afrique utile*, where superior profits could be made (Reno, 1999).

After independence, the Kenyan government continued to carry out certain aspects of the colonial agenda in northern Kenya, including promoting agriculture and sedentarizing pastoralists (Catley et al., 2013; Fratkin, 1997; Korf, Hagmann, & Emmenegger, 2015; Schrepfer & Caterina, 2014). Kenya's often-quoted first development strategy articulated the rationale for continuing with minimal state involvement in the north. It recommended that 'development money should be invested where it will yield the largest increase in net output', with the goal of 'favour[ing] the development of areas with abundant resources, good land and rainfall, transport and power facilities, and people receptive to and active in development' (RoK, 1965, p. 46). This approach to allocating public funds assumed that the benefits of agricultural development would eventually trickle down to benefit less productive regions of the country, reflecting a linear approach to rural development widely promoted by development economists of the time (Rostow, 1960).

In line with this understanding of rural development, the Kenyan government viewed pastoralism as a stage of socio-economic development that 'was expected to die a 'natural death' in response to modernization' (Idris, 2011, p. 25). Rather than seeking to develop and benefit from livestock trade, government intervention in the north aimed to speed up the process of sedentarization (Idris, 2011). A group ranching scheme was implemented in the 1960s by the Kenyan government, with expertise and funding from the World Bank and other bilateral donors (BurnSilver, 2009). Nomadic pastoralists were encouraged to settle in group ranches – 'privately titled collective rangelands used for communal livestock production' (Nelson, 2012, p. 3). Like other land titling schemes implemented in other parts of the world during this time, there was a clear 'improving' rationale for settling pastoralists and enclosing land (Hall, 2013; Li, 2007, 2010). Authorities claimed that group ranches would increase communal livestock production and improve land security, as well as facilitate the delivery of government services to pastoralist communities (Nelson, 2012).

By the mid-1970s, however, most group ranches in northern Kenya were failing to live-up to their intended objectives. The enclosed lands could not adequately support livestock and were degrading, sparking further conflict between pastoralists over resources (Nelson, 2012). Some group ranches were dissolved or subdivided because of such issues, while others were taken over by political elites (Sundstrom, 2009). The shortcomings of the group ranch scheme illustrate how previous attempts by government and development experts to reorganize land use and tenure arrangements and to limit the mobility of pastoralists produced unintended and often negative consequences for northern Kenya's population and ecology.

In more recent years, the relative absence of the national government from northern Kenya has created space for non-state actors to exert greater influence over land and resources in the region. Beginning in the mid-1990s, for example, some white settlers who have maintained large land holdings in the northcentral highlands identified an opportunity to incorporate land belonging to pastoralists into the conservation tourism industry (interview with representative of pastoralist civil society organization, Isiolo, May 2016). With support from international donors, these settlers-turned-conservationists encouraged pastoralists to consolidate land, remove fences, and create community conservancies in group ranches with the notion that this would benefit livestock and wildlife – as well as the general wellbeing of pastoralist communities.

Today, community conservation is expanding rapidly across northern Kenya. Although some pastoralist communities are active participants in conservation, others claim that conservancies are being established to benefit white settlers and political elites, while placing new constraints on access to land and resources (Bersaglio, 2017). In many ways, the issues emerging around community conservancies are reminiscent of earlier attempts to reconfigure space in northern Kenya, leading one civil society advocate to describe community conservation as 'Colonialism 2.0' (interview with representative of pastoralist civil society organization, Laikipia, May 2016). This sentiment aligns with existing critiques of East Africa's expanding conservation sector, which suggest that conservation has come to play a critical role in the consolidation of power and control over territory and resources by powerful state and non-state actors at the expense of relatively less powerful groups (Bersaglio, 2017; Gardner, 2012; German et al., 2016; Neumann, 2004).

The Lamu Port–South Sudan–Ethiopia Transport corridor

Northern Kenya's long history of contentious land politics is now being further complicated by recent discoveries of oil, geothermal and wind resources in the region and by the development of LAPSSET. Once complete, LAPSSET will include pipeline infrastructure to transport crude oil from northern Kenya's new oil fields to a new port on Kenya's coast. The corridor plan also includes a highway and railway network, international airports and a series of spatial development initiatives, including resort cities, an agricultural growth zone, and an export processing zone. In terms of land use planning, LAPSSET will be comprised of a 500-metre-wide corridor for transport infrastructure, overlaid by a 50-kilometre-wide economic corridor on either side for industrial and agricultural investment (LCDA, 2016). Laurance et al. (2015) estimate that the transport corridor will have a 1,500,000-hectare footprint once complete.

State agencies, as well as other authorities acting on behalf of the state, have used various acts of territorialization to facilitate the development of LAPSSET. First, a map of the corridor was produced that demarcated the route of the project, even prior to the discovery of oil in Kenya. At this time, the plan for LAPSSET included an oil pipeline running from South Sudan through Kenya to a port on the Kenyan coast. The original LAPSSET map was based on a feasibility study conducted by Japanese consultants that identified this as the best route for the transport corridor. As Bluwstein and Lund write, 'maps are imbued with authority and power not by accurately or "objectively" depicting territorialized spaces, but by actors seeking support for their claims through these maps' (2018, p. 456). In other words, although feasibility studies and map-making may appear to be a technical, objective and apolitical processes, such documents are used as visual representations to back territorial claims to land and resources.

After demarcating LAPSSET's route through northern Kenya, the LCDA – with the support of the National Land Commission (NLC) – pursued more overt political acts of physically claiming and acquiring land for the corridor. In July 2016, the LCDA proposed 'converting the required land along the corridor, whether it be private, community, or public land, to be 'Land Banked' so that access to this land is guaranteed as construction progresses' (LCDA, 2016, p. 38). During the latter part of 2016, the process of land acquisition began to intensify. The government issued a notice of intent to 'acquire 450 hectares for the Lamu resort, 81,811 hectares for the planned Lamu special economic zone, 10,744 hectares for the Lamu industrial zone, a further 28,500 hectares for the Lamu port, and 5,012 hectares for the Isiolo resort city' (Ngugi, 2016, par.5). In early 2018, the NLC confirmed that it would begin to acquire 197,200 hectares of land for all LAPSSET projects along the corridor (Kamau, 2018).

Since 2016, the LCDA has been deploying beacons and security installations along certain parts of the proposed corridor route to further restrict and secure land access (interview with civil society advocate, Laikipia, May 2016). In some cases, pastoralists have returned to their seasonal grazing areas only to find that beacons have been set up on the land without their knowledge. When this occurs, the beacons serve to inform pastoralists that land has been claimed for LAPSSET without their consultation (interview with civil society advocate, Laikipia, May 2016). In other cases, access roads for different components of the LAPSSET corridor, such as resort cities, have been cleared without formal consultation with pastoralists. Together, these steps are paving the way for investors, making territory and resources in northern Kenya more 'legible' – specifically, accessible, secure and sellable (Hall, 2013; Li, 2007; Scott, 1998).

Negotiating LAPSSET and reterritorializing land through rural resistance

This section examines the impacts of LAPSSET on northern Kenya's rural population, focusing broadly on two of the region's most predominant land users, pastoralists and conservationists. Specific strategies and tactics being used by these two groups to negotiate or resist LAPSSET related development are also examined. This section discusses the outcomes of such efforts and considers why these outcomes differ for pastoralists and conservationists respectively.

Pastoralists' reactions to LAPSSET in northcentral Kenya

The development of LAPSSET, along with the broader process of territorial restructuring taking place across northern Kenya, has unique implications for pastoralist groups. Pastoralists require access to large areas of land to sustain their livestock, but mega-infrastructure projects create new competition for land and other resources (Letai, 2015; Letai & Tiampati, 2014). Mobility is also essential to pastoralists, as most regularly move their livestock in search of pasture and water. This mobility is especially important during periods of drought or scarcity. Without adequate consideration and planning, mega-infrastructure projects can place new restrictions on the ability of pastoralists to migrate (Letai & Tiampati, 2014). Pastoralists also rely on healthy ecosystems. In addition to increased air and noise pollution due to more traffic, the potential for pipeline leaks creates a further layer of environmental risk for pastoralists.

LAPSSET raises important questions around land use, tenure, and value for pastoralists (Letai, 2015). Many pastoralists in northern Kenya live and graze their animals on untitled land, often referred to as community land. Pastoralists may have access and use rights to land, but there is often a lack of clarity around who formally 'owns' community land. In some cases, community land is held in trust by the government or political elites, which has made it possible for elites to sell off title deeds to community land for speculative purposes without consulting pastoralists (interview with civil society advocate, Laikipia, May 2016). Moreover, current land regulations make it possible for pastoralists to lose community land – including access to seasonal grazing lands or watering points – without monetary compensation if the land is needed for the 'common good' of the country, such as for the purpose of transport infrastructure (Letai & Tiampati, 2014).

Because LAPSSET carries risks for pastoralist livelihoods, networks of pastoralists and civil society organizations have been involved in negotiating and resisting components of the transport corridor. For the past several years, a coalition of pastoralist organizations has planned an annual 'Camel Caravan', which involves a multi-day march across northcentral Kenya that culminates on United Nations International Day of the World's Indigenous People. At one level, this demonstration serves as a platform to lobby the government to protect pastoralists against the negative livelihood impacts of proposed mega-infrastructure development in the region. For example, the caravan has drawn attention to the impacts of a mega-dam that has been proposed to service LAPSSET related projects (interview with civil society advocate, Laikipia, May 2016). However, pastoralists also use this demonstration as a strategy for demanding recognition for their territorial rights to land and resources as Indigenous peoples.

Beyond the Camel Caravan, networks of pastoralists and civil society organizations are also producing and publishing research that shows how land has been acquired for LAPSSET through processes and mechanisms that violate Indigenous peoples' rights (Sena, 2012; SWT, 2016; PDNK, 2016). This research provides documentation of cases in which LAPSSET construction began prior to informing and consulting communities, as well as cases where pastoralists have not been adequately compensated for losing access to land and resources because they lack formal title deeds. These networks have also monitored cases where rights guaranteed to Indigenous peoples by international law have not been respected during land acquisitions. In response to some of this research, the United Nations Permanent Forum on Indigenous Issues undertook an unofficial mission with the support of the International Working Group on Indigenous Affairs (IWGIA), to explore the situation of Indigenous Peoples along the LAPSSET corridor in Kenya.

Finally, pastoralist networks and organizations have also attempted to create their own rules and regulations to further solidify their control over land and resources in northern Kenya. For example, a non-governmental organization in the coastal region of northern Kenya, Save Lamu, has developed a Biocultural Community Protocol (BCP), which articulates community-determined values, procedures and priorities for accessing and managing land and resources (Natural Justice, 2019). The BCP documents the procedures that the government and investors must follow before proceeding with any development in the region to comply with local, national and international law. To create the BCP, Save Lamu engaged with communities, including pastoralists, in mapping their resources and documenting their vision for land and resources use.

These examples demonstrate how pastoralists are attempting to draw territorial boundaries from below through acts of reterritorialization (or re-re-territorialization) to challenge the state's territorial restructuring from above. These acts are often quite overtly political, as the government does not officially recognize any ethnic groups within its borders as Indigenous. By self-identifying as Indigenous, aligning with global Indigenous peoples' movements and demanding that the state respect international conventions on Indigenous rights, pastoralists are forging global alliances and using global legal discourse to exert pressure on the state to administer land and natural resource development in a way that complies with global norms (Tarrow, 2011).

The efforts of pastoralist networks to exert authority over land and resources in the region have caught the attention of the Kenyan government. The government has created a steering committee to liaise between communities and government officials about LAPSSET construction (interview with civil society advocate, Laikipia, May 2016). Pastoralist organizations and leaders have also been invited to inform political discussions about compensation frameworks and resettlement plans for displaced communities (interview with civil society advocate, Laikipia, May 2016; interview with civil society advocate B, Laikipia, May 2016). Finally, the recently published *Strategic Environmental Assessment for the LAPSSET Corridor* recognizes pastoralists as Indigenous peoples that stand to be uniquely impacted by the project and commits to formulating and implementing a mitigation strategy based on the concerns of pastoralist communities (LCDA & Repcon Associates, 2017). This suggests that the state-led process of territorial restructuring in northern Kenya can be influenced by the territorial claims of the region's rural land users. This reinforces other recent work which has found that pastoralists in northern Kenya have had some success in expressing agency over their future amid rapidly changing land dynamics in the region (Catley et al., 2013; Korf, et al., 2015).

Conservationists' reactions to LAPSSET in northcentral Kenya

Conservation is another dominant land use in northern Kenya. The region is home to over 75 per cent of the country's wildlife, along with 18 national parks, reserves and sanctuaries and 33 community conservancies (Little, McPeak, Barrett, & Kristjanson, 2008; NRT, 2015). These spaces support much of Kenya's biodiversity and serve as key migratory routes for wild animals. Northern Kenya also contains habitat for endangered and critically endangered species of wildlife, such as black rhino, Cape hunting dog, elephant, Grevy's zebra, hirola, and reticulated giraffe. Beyond protecting these and other species, the wildlife sector contributes to the region's local economy and the national economy as well. The wildlife sector includes for-profit and not-for-profit industries associated with wildlife conservation, ranging from eco- and nature tourism to ecosystem or species research. For example, northern Kenya's largest conservation organization, the Northern Rangelands Trust (NRT), claims to provide some degree of economic support to over 250,000 people across the region on an annual basis (NRT, 2015).

Initial proposals suggested that LAPSSET would bisect or run parallel to internationally protected heritage sites and world-renowned conservation areas, such as Lewa Wildlife Conservancy and Samburu National Reserve. The impacts of this transport corridor on conservation efforts in northern Kenya stand to be lasting and potentially harmful. The construction and operation of new highways, access roads, rail lines, and pipelines risks disrupting these spaces as well as wildlife and human or livestock migratory patterns. Given the potential for high environmental costs and habitat destruction some conservation organizations have gone so far as to claim that the costs of LAPSSET outweigh its benefits (interview with representative of conservation organization, Laikipia, May 2016). They also argue that components of the project are 'clearly incompatible' with national and international objectives of biodiversity conservation (interview with representative of conservation organization, Laikipia, May 2016).

In addition to the direct environmental impacts of the transport corridor, LAPSSET also presents significant environmental risks from a conservation perspective. For example, the original sites proposed for LAPSSET's three resort cities are in complex socio-ecological areas. Isiolo resort city was originally planned for a remote area of northcentral

Kenya that has traditionally served as a source of dry-season pasture for pastoralists, as well as an important corridor for elephants that migrate to and from Samburu County to the Aberdare Range and/or Mt. Kenya (interview with community leader, Isiolo, May 2016). The construction of a dam on the Ewaso Ng'iro River near Oldonyiro has been proposed to service this resort city, as well as to provide irrigation for new agricultural projects located both down- and upstream from the dam. The proposed dam reservoir would submerge a total of 2,083 hectares of conservancy land, requiring additional land to be sequestered for the dam's power station and switchboard (SWT, 2016). Submerging this land upstream of the dam would place new pressures on land and livelihoods on both sides of the reservoir – impacting humans, livestock and wildlife, as well as the ecological relationships in which they are all embedded (interview with representative of conservation organization, Laikipia, May 2016; SWT, 2016).

Recognizing the various environmental risks that LAPSSET presents, conservation actors in the region are undertaking efforts to negotiate the terms of the transport infrastructure and to resist components of the project. One group at the forefront of this movement is a coalition of national and international conservation and civil society organizations working together as the Ewaso Ng'iro Basin Stakeholder Forum (ENBSF). This Forum was established to share information about the environmental impacts of LAPSSET and to spearhead a coordinated response to the project. The ENBSF includes representatives from pastoralists advocacy networks and organizations. Yet the strategies, tactics and outcomes being pursued by the group arguably privilege the interests of conservationists over pastoralists – although in some spaces the interests of both groups complement each other or overlap. The ENBSF has conducted an integrated technical review of components of the LAPSSET corridor, engaging economists, ecologists, hydrologists, civil engineers and resettlement experts in the process. Similarly, individual conservation organizations have developed maps, which portray ecosystem services and human and wildlife activities in relation to proposed LAPSSET developments. These studies have been used to help conservationists quantify the environmental costs they attribute to LAPSSET – ranging from lost tourism revenues to impacts on wildlife mobility – to ensure that these costs are considered in the government's cost-benefit analysis of the project (interview with representative of conservation organization, Nairobi, May 2016).

Some conservation organizations have also raised concerns about the legal implications of moving ahead with LAPSSET without fully understanding the environmental impacts of the project and developing appropriate mitigation measures in response. Construction on components of LAPSSET is underway; however, many conservation organizations question whether national environmental regulations were upheld before commencing the project (Interview with representative of conservation organization,

Nairobi, May 2016; interview with representative of conservation organization, Laikipia, May 2016). Using their own technical studies to justify the need for further environmental research, conservation organizations have called on the state to carry out more comprehensive environmental assessments before moving forward with the project. Presumably in response, LAPSSET authorities hired consultants to carry out a Strategic Environmental Assessment (SEA) for the project in 2016.

Conservation networks negotiating over LAPSSET are comprised of heterogeneous actors, including: white settlers and foreign nationals; conservation tourism investors, managers and employees; international organizations; and national and international scientists, among others. Like pastoralist networks and organizations, conservationists are also exercising territorial claims to land and resources in response to LAPSSET. Some of these claims are rooted in cultural identity, as some white settlers have aligned with conservationists to defend their properties from LAPSSET and other changing land dynamics in the region (Bersaglio, 2017). However, territorial claims by conservationists have more often been rooted in ideas about efficiency and capacity – in other words, claims about who is best fit to manage the region's landscape and wildlife resources. By challenging the adequacy of official environmental plans and studies conducted for LAPSSET, conservation actors have reasserted their collective authority as those best-suited to exercise control over northern Kenya's complex socio-ecological landscape and valuable biodiversity. This is not mere sentiment. Kenya's new Wildlife Conservation and Management Act of 2013 devolves authority and responsibility for wildlife conservation to non-state actors – including private landowners and the private sector.

The state has responded to the wildlife sector's mobilization to exert authority over how land should be used, by whom, and to what end in northern Kenya. In late 2016, numerous news sources in Kenya reported that local politicians in northern Kenya were demanding that the LCDA conduct further studies to better understand the impact of LAPSSET on the conservation sector. The LCDA's *Strategic Environmental Assessment for the LAPSSET Corridor* published in January 2017 responded to these demands, outlining how the potential impacts of LAPSSET on northern Kenya's valuable wildlife resources will be mitigated in an 'Environmental and Social Management Plan'. Most importantly, the 'Environmental and Social Management Plan' recommended rerouting the corridor and relocating the resort city to bypass areas that are important for certain wildlife species, such as elephants.

The uneven outcomes of reactions to LAPSSET in northcentral Kenya

Despite achieving some collective success in negotiating and resisting LAPSSET, the ambitions, expectations, and interests that have motivated reactions on the part of pastoralists and conservationists are not in complete alignment. Key points of divergence can be traced back to efforts by colonial and post-colonial governments to reorganize space in northern Kenya to privilege white settlement, commercial ranching and, importantly, conservation. Therefore, although pastoralists and conservationists find themselves in similar struggles for territory as a result of natural resource and infrastructure development in the north, they struggle from very different positions in society. This has real implications for the strategies available to both groups in response to LAPSSET, as well as for the types of outcomes that are desired and possible.

Just as the creation of a 'white reserve' in Kenya's highlands by the colonial administration set in motion many of the structural forms of violence that continue to disadvantage pastoralist groups in northern Kenya today, so too did the creation of game reserves and conservation areas that followed in the 1990s. Today, the rapid expansion of conservancies on lands belonging to pastoralists continues. In recent years, the NRT has expanded its presence in northern Kenya to such an extent that some pastoralists now refer to the organization as 'the new government' in the north (interview with representative of pastoralist civil society organization, Laikipia, May 2016). With this in mind, pastoralists and conservationists are struggling to reclaim territory in northern Kenya in response to LAPSSET on uneven terrain.

Pastoralists and conservationists are also motivated by different objectives in their responses to LAPSSET. Pastoralist networks and organizations involved in negotiating or resisting LAPSSET are often trying to reinforce and defend the integrity of their territories amid ever-shrinking space for pastoralism. Such efforts are largely motivated by the material needs of pastoralists – including access to sufficient pasture and water reserves – as well as by cultural notions of livelihood, territory and wellbeing. As one pastoralist civil society organization has argued, access to and control over 'land is not just the means for economic survival but also the basis of their cultural identity and spiritual wellbeing' (PDNK, 2016, p. 14). Redressing the persistent legacies of colonial dispossession through resistance to new large-scale investments and development is also a driving motivation for pastoralists. This does not mean that all pastoralists oppose all aspects of LAPSSET, as many individuals and communities see some potential to benefit from new infrastructure as well as from access to new employment opportunities and markets associated with the corridor.

In contrast, conservationists are trying to defend land boundaries demarcated for conservation tourism and to maintain the integrity of wildlife habitat and migratory corridors in the process. Efforts to negotiate or resist LAPSSET on the part of conservationists are, therefore, motivated by the material needs of the wildlife sector – including land, ecosystems and migratory routes for wildlife – as well as by cultural notions and imaginaries of what 'African' landscapes should look like and how they should be managed. Unlike pastoralists, conservationists are generally unconcerned with redressing the legacies of colonial dispossession. In fact, efforts to restructure territory on the part of conservationists sometimes flies directly in the face of this objective. For example, representatives of pastoralist and conservation groups alike reported that some white landowners near Oldonyiro welcome the construction of the dam on the Ewaso Ng'iro as part of the LAPSSET corridor. Not only would the dam afford them new conservation tourism opportunities, such as boat safaris and sport fishing, but it would serve as a natural barrier against pastoralists trying to illegally graze and water their livestock on white-owned properties (interview with representatives of conservation civil society organization, Nairobi, May 2016).

In line with the conflicts and inequalities that historically define social relationships between pastoralists and conservationists in northern Kenya, the two groups have faced challenges in mounting a unified resistance against the state-led process of territorial restructuring. Although pastoralists and conservationists are involved in some of the same civil society networks mobilizing in response to LAPSSET, the interests of these groups are not always equally represented in these networks. For example, when local media has reported on civil society events or workshops pertaining to the impacts of LAPSSET, issues around wildlife tourism and biodiversity conservation have been more readily reported on than issues around land tenure and compensation for pastoralists.

The outcomes of efforts to negotiate LAPSSET and to reterritorialize land through rural resistance seem to weigh in favour of conservationists as well. Conservationists have exerted tangible influence over the direction of LAPSSET-related developments. The most evident example of this influence is that certain components of the transport corridor have been stalled or re-routed in response to conservationists' concerns about ecosystem health, wildlife corridors and conservation tourism. This newly proposed route for LAPSSET still passes through areas important to pastoralists, but is now further away from private wildlife areas. The disparity in the ability of pastoralists and conservationists to influence LAPSSET does not reflect a lack of agency among pastoralists; rather, due to the privileged position that conservationists occupy in the north, they have so far collectively achieved more advances in reterritorializing land from below in response to LAPSSET.

Conclusion

This chapter contributes to research on rural social movements by demonstrating how rural groups can use various strategies to negotiate the state's claims to territory and resist state-led efforts to reclaim territory for natural resource development. It is argued that, in the case of northern Kenya, rural groups are engaging in acts of reterritorialization 'from below' in response to the state's efforts to make land and natural resources more amenable to investment 'from above'. The analysis carried out in this chapter also provides insights into the different and sometimes competing notions of territorial rights that animate resistance to natural resource development in rural society. By analysing how networks of pastoralists and conservationists have reacted to LAPSSET thus far, as well as the outcomes such groups have achieved respectively, this chapter demonstrates that rural groups resisting natural resource development are often driven by diverse motives and that their resistance can have uneven outcomes.

This chapter also draws attention to the various forms of differentiation that permeate rural social movements. Pastoralists and conservationists have independently and collectively achieved important successes in negotiating the terms of LAPSSET since construction began in 2012. Pastoralist networks have succeeded in shaping discourses and policies about land acquisition for LAPSSET, ensuring that, in the very least, some consideration is afforded to their unique concerns and interests. Conservationists, however, have shaped the planning of LAPSSET in more obvious ways. By reasserting territorial claims to land and resources in Kenya's north, conservation networks have achieved some success in reclaiming space for conservation tourism, wildlife habitat and migratory corridors that were originally allocated for LAPSSET development. Importantly, because the final routing of LAPSSET is yet to be determined, the staying power of resistance is also yet to be determined.

The differentiated ability of pastoralists and conservationists to 'modify, subvert, block and even overturn' LAPSSET reflects the different types of social differentiation and stratification that underpin rural society in northern Kenya. In this sense, the strategies, tactics and outcomes of resistance pursued by different rural groups have been shaped – to various extents – by historical trajectories of class, ethnic and racial relationships, as well as by the spatial configurations of these and other social relationships in the landscape. Although this chapter demonstrates that acts of reterritorialization can be successful in resisting natural resource development, it also reveals that whether or not this approach to resistance is successful is shaped by existing inequalities and points of contention within rural society. With this in mind, research on rural social movements should not only consider how and why different rural groups react to natural resource development but also the uneven social terrain on which different groups struggle over divergent notions of territory.

References

Alonso-Fradejas, A. (2015). Anything but a story foretold: Multiple politics of resistance to the agrarian extractivist project in Guatemala. *Journal of Peasant Studies*, 42, 3–4, 489–515.

Bersaglio, B. (2017). *Safari tourism, the green economy and 'acquiescence' in Laikipia, Kenya: A genealogy of green grab.* Doctoral dissertation. University of Toronto.

Bluwstein, J. & Lund, J. F. (2018). Territoriality by conservation in the Selous–Niassa corridor in Tanzania. *World Development*, 101, 453–465.

Borras Jr, S. M. & Franco, J. C. (2013). Global land grabbing and political reactions 'from below'. *Third World Quarterly* 34, 9, 1723–1747.

Borras Jr, S. M., Hall, R., Scoones, I., White, B., & Wolford, W. (2011). Towards a better understanding of global land grabbing: An editorial introduction. *The Journal of Peasant Studies* 38, 2, 209–216.

Brent, Z. (2015). Territorial restructuring and resistance in Argentina. *The Journal of Peasant Studies*, 42, 3–4, 671–694.

BurnSilver, S. B. (2009). Pathways of continuity and change: Maasai livelihoods in Amboseli, Kajiado District, Kenya. In K. Homewood, P. Kristjanson, & P. Trench, (Eds) *Staying Maasai? Livelihoods, Conservation and Development in East African Rangelands (Vol. 5)* (pp. 161–207). New York: Springer.

Catley, A., Lind, J., & Scoones, I. (2013). *Pastoralism and Development in Africa: Dynamic Change at the Margins.* New York: Routledge.

Cavanagh, C. J. & Benjaminsen, T. A. (2015). Guerrilla agriculture? A biopolitical guide to illicit cultivation within an IUCN Category II protected area. *Journal of Peasant Studies* 42, 3–4, 725–745.

Corson, C. (2012). Territorialization, enclosure and neoliberalism: non-state influence in struggles over Madagascar's forests. *The Journal of Peasant Studies*, 38, 4, 703–726.

Deininger, K. (2011). Challenges posed by the new wave of farmland investment. *The Journal of Peasant Studies* 38, 2, 217–247.

Eriksen, S. & Lind, J. (2009). Adaptation as a political process: Adjusting to drought and conflict in Kenya's drylands. *Environmental Management* 43, 5, 817–835.

Fratkin, E. (1997). Pastoralism: Governance and development issues. *Annual Review of Anthropology*, 235–261.

Gardner, B. (2012). Tourism and the politics of the global land grab in Tanzania: markets, appropriation and recognition. *Journal of Peasant Studies* 39, 2, 377–402.

German, L. A., Unks, R., & King, E. (2017). Green appropriations through shifting contours of authority and property on a pastoralist commons. *The Journal of Peasant Studies*, 44, 3, 631–657.

Greiner, C. (2016). Land-use change, territorial restructuring, and economies of anticipation in dryland Kenya. *Journal of Eastern African Studies*, 10, 3, 530–547.

Hall, D. (2013). *Land.* Cambridge: Polity.

Hughes, L. (2006). *Moving the Maasai: A Colonial Misadventure.* Hampshire: Palgrave Macmillan.

Idris, A. (2011). Taking the Camel through the Eye of a Needle: Enhancing Pastoral Resilience through Education Policy in Kenya. *Resilience: Interdisciplinary Perspectives on Science and Humanitarianism*, 2, 25–38.

Kamau, M. (2018, 13 January). Land commission to start acquisition half a million acres for LAPSSET projects. *The Standard.*

Korf, B., Hagmann, T., & Emmenegger, R. (2015). Respacing African drylands: territorialisation, sedentarisation and indigenous commodification in the Ethiopian pastoral frontier. *Journal of Peasant Studies* 42, 5, 881–901.

LCDA. (2016, July). *Brief on LAPSSET Corridor Project*. Nairobi: Lapsset Corridor Development Authority.

LCDA & Repcon Associates. (2017 January). *Strategic Environmental Assessment Study in the Master Plan for the LAPSSET Corridor Infrastructure Development Project (LCIDP) – Draft Report*. Nairobi: Lapsset Corridor Development Authority.

Laurance, W. F., Sloan, S., Weng, L., & Sayer, J. A. (2015). Estimating the environmental costs of Africa's massive 'development corridors'. *Current Biology* 25, 24, 3202–3208.

Letai, J. (2015). *Role of community land in the development of the ASALS in Kenya.* Paper presented at Reconcile Conference on Land and Natural Resources, Sarova Panafric Hotel, Nairobi, Kenya, 12 March 2015.

Letai, J. & Tiampati, M. (2014). Capturing Benefits Whilst Safeguarding Livelihoods: The Debate over LAPSSET. In H. de Jode and V. Tilstone (Eds) *Resilience in the Drylands of the Horn of Africa – Edition 5: Experiences and Lessons Learnt for Improved Policy and Practice*, (pp. 30–32). Nairobi: Drylands Learning and Capacity Building Initiative (DLCI).

Li, T. M. (2007). *The Will to Improve: Governmentality, Development, and the Practice of Politics*. Durham, NC: Duke University Press.

Li, T. M. (2010). To make live or let die? Rural dispossession and the protection of surplus populations. *Antipode* 41, s1, 66–93.

Little, P. D., McPeak, J., Barrett, C. B., & Kristjanson, P. (2008). Challenging orthodoxies: Understanding poverty in pastoral areas of East Africa. *Development and Change* 39, 4, 587–611.

Lund, C. (2011). Fragmented sovereignty: Land reform and dispossession in Laos. *The Journal of Peasant Studies*, 38, 4, 885–905.

Margulis, M., McKeon, N., & Borras Jr., S. (2013). Land grabbing and global governance: critical perspectives. *Globalizations*, 10, 1, 1–23.

Natural Justice. (2019). Kenya's Lamu communities launch a united Biocultural Community Protocol. Retrieved 24 June 2019 from https://naturaljustice.org/kenyas-lamu-communities-launch-a-united-biocultural-community-protocol/. Accessed 24 June 2019.

Nelson, F. (2012). *Community Rights, Conservation and Contested Land: The Politics of Natural Resource Governance in Africa*. New York: Routledge.

Neumann, R. P. (2004). Nature-State-Territory: Toward a critical theorization of conservation enclosures. In R. Peet and M. Watts (Eds) *Liberation Ecologies: Environment, Development and Social Movements*, (pp. 179–196). New York: Routledge.

New Humanitarian. (2013). Pastoralism's economic contributions are significant but overlooked. 6 May 2013. Nairobi: The New Humanitarian (formally IRIN News). Accessed December 2016 at: www.thenewhumanitarian.org/news/2013/05/16/pastoralism-s-economic-contributions-are-significant-overlooked.

Ngugi, B. (2016). Swazuri team plans Lapsset land payouts. *Business Daily*, 23 October 2016.

NRT. (2015). *NRT State of the Conservancies Report 2014*. Lewa: Northern Rangelands Trust.

PDNK. (2016). LAPSSET: *Voices from the Ground*. Nairobi: Pastoralist Development Network of Kenya.

Peluso, N. (2005). Seeing property in land use: Local territorializations in West Kalimantan, Indonesia. *Geografisk Tidsskrift, Danish Journal of Geography*, 105, 1, 1–15.

Peluso, N. & Lund, C. (2011). New frontiers of land control: Introduction. *The Journal of Peasant Studies*, 38, 4, 667–681.

Reno, W. (1999). *Warlord Politics and African States*. Boulder, CO: Lynne Rienner Publishers.

RoK – Republic of Kenya. (1965). African socialism and its application to planning Kenya. *Sessional Paper Number 10*, Nairobi.

Ribot, J. & Peluso, N. (2003). A theory of access. *Rural Sociology*, 68, 2, 153–181.

Rocheleau, D. (2015). Networked, rooted and territorial: green grabbing and resistance in Chiapas. *The Journal of Peasant Studies*, 42, 3–4, 695–723.

Rostow, W. W. (1960). *The stages of growth: A non-communist manifesto*. Cambridge: Cambridge University Press.

Salih, M. A. Mohamed, T. Dietz, & Abdel Ghaffar Mohamed Ahmed. (2001). *African Pastoralism: Conflict, Institutions and Government*. London: Pluto Press.

SWT. (2016). *Report on Indigenous Women and the Lapsset Corridor in Laikipia and Isiolo Counties*. Laikipia: Samburu Women's Trust.

Schrepfer, N. & Caterina, M. (2014). *On the margin: Kenya's Pastoralists – from displacement to solutions, a conceptual study on the internal displacement of pastoralists*. Geneva: Internal Displacement Monitoring Centre.

Scott, J. (1998). *Seeing Like a State: How Certain Schemes to Improve the Human Condition have Failed*. New Haven, CT: Yale University Press.

Sena, K. (2012). *Lamu Port–South Sudan–Ethiopia Transport Corridor (LAPSSET) and Indigenous Peoples in Kenya*. Copenhagen: IWGIA – International Working Group on Indigenous Affairs.

Steinhart, E. (2006). *Black Poachers, White Hunters: A Social History of Hunting in Colonial Kenya*. Athens, OH: Ohio University Press.

Sundstrom, S. (2009). *Rangeland privatisation and the Maasai experience: implications for livestock herding, open space, and wildlife conservation in southern Kenya*. Master's Thesis, Oregon State University.

Tarrow, S. (2011). *Power in Movement: Social Movements and Contentious Politics, 3rd edition*. Cambridge: Cambridge University Press.

Vandergeest, P. & Peluso, N. (1995). Territorialization and state power in Thailand. *Theory and Society*, 24, 3, 385–426.

4 Confronting neoliberal resource policy

Mining conflict and coal politics in Bangladesh

M. Omar Faruque

Introduction

Many resource rich countries in the global periphery fail to generate positive economic growth from natural resource development, a phenomenon known as the 'resource curse' (Collier, 2010; Humphreys et al., 2007). On the one hand, extractive capital and the state seek resource extraction for profit. On the other hand, local communities try to stop energy and mining corporations from growing in the global periphery (Bebbington, 2012; Chowdhury, 2016; Deonandan & Dougherty, 2016; Horowitz & Watts, 2017). These incompatible goals inevitably lead to political conflict (Le Billon, 2012).

Mining corporations frequently underestimate the social, political, and environmental risks of their operations in host communities (Butler, 2015; Kirsch, 2014; Li, 2015). Power elites, driven by a neoliberal agenda, pay little attention to risks as they are preoccupied with short-term economic gains. Local communities, who bear the brunt of mining operations, think otherwise. They offer counter-narratives based on their lived experience (Bebbington & Burry, 2013; Luthfa, 2011; Macdonald et al., 2017; Nuremowla, 2016; Singh & Camba, 2016), successfully attracting civil society actors to their struggle against resource development (Kumar, 2014; Ozen & Ozen, 2017). Mining corporations respond by inventing corporate social responsibility (CSR) programmes to minimize discontent in host communities (Gardner, 2012). Although the ostensible goal of these programmes is to improve corporation-community relations, their real objective is to delegitimize discontent (Horowitz, 2015). As several scholars have shown, CSR programmes lack genuine interest in the wellbeing of host communities; this new corporate practice is designed to secure (and maximize) profit (Rajak, 2011; Walker-Said & Kelly, 2015).

Anti-mining movements confront other powerful forces as well. For example, international financial institutions such as the World Bank and the Asian Development Bank push for neoliberal economic transformation to create opportunities for foreign capital in the periphery (Muhammad, 2014a), supporting energy and mining companies through political risk insurance loans. Although political risk insurance is a financial instrument, it has serious political ramifications (Moody, 2005). Western development agencies working in the global South such as USAID and DFID play a significant role in the institutionalization of neoliberal reforms to maximize economic opportunities for multinational corporations (Muhammad, 2003). They support poverty alleviation programmes, which are closely linked to their corporate business strategies. Finally, national and international NGOs often act as partners of mining companies to safeguard corporate interests. These actors are an integral part of a symbiotic power complex seeking to institutionalize neoliberal resource policy in the global South.

This chapter analyses the anti-mining movement confronting neoliberal resource policy in Bangladesh. A decade-long political struggle against a British mining corporation, GCM Resources plc, and its wholly-owned subsidiary, Asia Energy Corporation (Bangladesh) Pty Ltd (AEC) serves as the empirical case study. This struggle had two intertwined scales of resistance: local and national. Although its broader agenda was to challenge the dominant paradigm of development influenced by the market calculus of neoliberal globalization (Muhammad, 2013, 2014b; also see McMichael, 2010), two specific contentious issues were at the centre. At the local scale, the key issue was the construction of a large open pit coal mine in Phulbari, a rural town 300 kilometres northwest of Dhaka; at the national scale, the main issue was the making of national coal policy. The National Committee to Protect Oil, Gas, Mineral Resources, Power and Ports (NCBD), a radical social movement organization, worked at the national level to strengthen local mobilization. Drawing on social movement scholarship (Giugni, et al., 1999; Klandermans, 1997), I consider mobilization strategies and movement outcomes to show how social movements can persuade powerful authorities to rethink their policy choices. Anthropologist Stuart Kirsch's (2014) theorizing of political struggles against mining corporations, particularly his conceptualization of 'politics of time' and 'politics of scale', is the main analytical framework.

Kirsch (2014) argues that political movements against mining corporations can realize their objectives if they deploy a new tactic, the politics of time, along with an old tactic, the politics of scale. The latter tactic includes scaling up mobilization efforts, with local activists forming strategic alliances with national and/or transnational advocacy groups to strengthen their agenda against powerful opponents. The politics of time is a new form of activism that seeks to confront mining operation before it begins. It aims to create obstacles so that mining corporations cannot easily mobilize capital and get government approval. It offers social movements the opportunity to confront 'the means by which elites extend their power over the body politic through their control over the social construction of time' (Kirsch, 2014, p. 191). The literature on mining conflicts shows that protests have little effect once production begins (Bebbington, 2012). Even with a well-organized movement including local, national, and transnational advocacy networks (i.e. politics of scale), affected communities can, at best, demand some compensation (Kirsch, 2014). They will not be able to remedy rights violations or to halt the 'slow violence' unleashed by neoliberal global capitalism. As Kirsch argues:

> The politics of time is ... central to contemporary environmental debates, especially the contradiction between the short term interests of capital and corporations and the longue durée of industrial impacts on the environment. New strategies invoking the politics of time must bring these longer temporal horizons into public consciousness and make them subject to political and economic calculations. This requires paying greater attention to the production of environmental risks and hazards rather than deferring their diagnosis and solution to the future.
>
> (2014, p. 191)

In short, the politics of time, when accompanied by the politics of scale, has enormous potential to confront mining corporations and prevent the dispossession of lands and livelihoods and environmental destruction; Kirsch substantiates his argument by pointing to the successes of several Latin American anti-mining movements which adopted this tactic.

The chapter focuses on the emergence and development of the Phulbari movement in Bangladesh; it considers the local dynamics of the movement and its larger collaboration with the national-level NCBD. It shows how local and national mobilization collectively resisted the neoliberal rationality of extractive capital and the state in a decade-long struggle (2005–2016) over the Phulbari coal mine and the national coal policy. Although the mining company and power elites did their best to ignore the counter-narratives of the challengers, the incessant activism finally persuaded the Bangladeshi government to acknowledge the merits of their agenda. The analysis draws attention to the significance of both 'politics of time' and 'politics of scale' in mobilizing a particular subaltern community to challenge powerful authorities.

Qualitative data, used in the chapter, come from multiple primary sources, including in-depth interviews with local and national activists and local people of various categories (peasants, minority *adivasis* [indigenous peoples], petty traders, agricultural labourers, businessmen, elected representatives, and community elites). Data also come from reports of the mining company and government agencies, media interviews of the officials of the Energy Ministry and the mining company, a set of organizational documents of various groups involved in the Phulbari movement, and newspaper reports on coal mining and coal policy for 2004–2016.

The chapter has two major sections. Section one analyses local resistance, looking at the anti-mining movement in Phulbari during 2005–2006. In the section, I consider the emergence, actions and processes, and outcomes of this local mobilization. I emphasize the historical and political constellations shaping settlement history and community relations. I also note some significant developments post-2006. Section two looks at the national level struggle over coal policy. In the section, I show how various antagonistic forces contested coal policy and the net outcome of this struggle. The chapter concludes with a brief note on the conditions and strategies making the local and the national movements exemplars of success.

Local community mobilization in Phulbari

In October 2005, Asia Energy Corporation (Bangladesh) Pty Ltd (AEC), a wholly owned subsidiary of a British mining company, GCM Resources plc (GCM, 2016), applied for government approval to build a large open pit coal mine in Phulbari. The mining company planned to acquire more than 6,000 hectares of land (AEC, 2005). It would produce 15 million tons of coal annually, exporting twelve million tons and keeping three million tons for the domestic market. The Bangladeshi government (GOB) would receive 6 per cent royalty. The mining company estimated that the coal mine would 'transform' the regional economy and significantly contribute to national economic growth by adding 1 per cent to annual GDP. It said the Bangladeshi government would also benefit from corporate tax and other developmental activities. It argued the regional economy would be transformed from 'subsistence' agricultural production to an industrial production system. Its feasibility study showed the coal mine would not cause any substantial risks for agriculture or the environment (particularly underground water resources). The net loss of agriculture would be zero and the changes in underground water resources were 'manageable' (AEC, 2005). The coal mine would require the displacement of more than 40,000 people, including 2,300 *adivasis* (AEC, 2006). Power elites (local and national) largely agreed with this economic narrative of the mining company and supported its agenda. An official of the Energy Ministry explained the significance of the project in the following way:

> Because of this coal [mine], there will be many industries and jobs will be created. I think that it will play an important role in the local economy. And so far as I can predict, this area might become the most economically viable region in the near future. There are so many natural resources here.[1]

Community leaders and those at the grassroots level were unconvinced by this economic narrative.[2] They said open pit mining was not suitable for a densely populated region like Phulbari with more than 850 people living per square kilometre (BBS, 2014). The mining method, they argued, would destroy agricultural land – the main source of their livelihood. Regional business leaders[3] noted that the region was dependent on agricultural production, with surplus crops delivered annually to the domestic food market (DAE, 2017). Local people also rejected the estimation of the displaced population. As they saw it, the mining company had reported only a fraction of the population who would be evicted.[4] According to local estimates, more than 500,000 people would be affected, and a vast area would suffer from a lack of water resources.[5] Official statistics corroborate these estimates (BBS, 2014). More significantly, local communities anticipated the loss of land, culture, heritage, and community cohesion. They did not support the mining company's plan to offer them financial compensation. The complex social and cultural relationship between land and community in rural Bangladesh, they argued, was beyond compensation.[6]

During community consultations, the mining company failed to satisfactorily explain its 'mine development scheme',[7] and the initial discontent became a well-organized anti-mining movement in June 2005. Community leaders urged local people to support the anti-mining agenda of the Committee to Protect Phulbari (CPP). An elected representative of Phulbari Municipality, the epicentre of the movement, said:

> In the beginning of 2005, Asia Energy communicated with local elders, elected leaders and conducted consultations. They didn't elaborate about environmental problems that can be caused by the project. They told us that it will be an open pit mining, but we didn't understand what open pit mining was. Then, concerned people of Phulbari formed a study group and they realized how destructive open pit mining could be. Then, we realized that there is a need to fight against this project and created a movement.[8]

CPP used various non-violent protest actions. These included petitions to the Prime Minister, rallies, human chains, weekly processions, press conferences, and *hartal*.[9] Its discourse centred on the fear of losing land, culture, memory, ancestral ties, and community (CPP, 2005a). It urged the government to cancel the project. Its petition to the Prime Minister expressing determination to resist the mining company, reads: 'We will not give our land for the so-called coal mine, which would benefit the mining company more than our country. We will sacrifice our lives, but will not allow anyone to take our forefather's graveyards' (CPP, 2005b).

CPP soon began to collaborate with the National Committee to Protect Oil, Gas, Mineral Resources, Power and Ports (NCBD), an organization well known for its anti-imperialist political struggle against 'neoliberal energy imperialism' in Bangladesh (Ahmed, 2013; Muhammad, 2004; Shaheedullah, 2002, 2009). NCBD diagnosed the project as 'resource plunder'. It articulated three key demands (three NOs): no open pit mining, no involvement of foreign mining company, and no export of mineral resources (NCBD, 2005).

In March 2006, CPP and NCBD jointly announced 'the Phulbari Declaration' at a rally. The declaration reads:

> If the project is implemented the coalmine will become AEC's property, a small portion ... will be offered to Bangladesh ... The open pit mining method will result in destruction of a prosperous area ... the cessation of all agricultural and other economic activities ... loss of archaeological treasures, including eviction of *lakhs* of people, and desertification of a vast area ... pollution of rivers, canals and wetlands in the vicinity. Those who attempt to portray this project of destruction and plunder as 'development,' and propagate the view that foreign investments are essential ingredients of 'development,' are committing a crime.
>
> (Muhammad, 2006, Par. 15)

Although CPP and NCBD continued the campaign both locally and nationally, neither the Bangladeshi government nor the mining company responded. Their disinterest forced NCBD (now the de facto leader of the movement) to change mobilization tactics. It announced a protest event that would *gherao* (surround) the mining company office in Phulbari on 26 August 2006. The protest turned violent, with state security forces killing three protesters. Not surprisingly, local people became even more vocal. As a result, the structural power tilted towards them; the government accepted all demands and signed a memorandum of agreement with the protesters. The first clause reads:

> All agreements executed with [the mining company] shall be scrapped and the company shall be driven out from the four [Upazila] including Phulbari and finally from [Bangladesh]. Open pit mining method for production of coal shall not be employed in four [Upazila] including Phulbari and, for that matter, anywhere in Bangladesh. Other method(s) to be employed for coal production shall be subject to people's consent.

Two major political parties (ruling and opposition parties in 2006) supported the memorandum. In this instance, the government went further; an expert committee, commissioned by the Energy Ministry, submitted its assessment of the 'mine development scheme' of the mining company in September 2006. The committee said: 'We have found there is no legal, environmental, economic, and technological basis to allow the company to mine and extract coal from the [Phulbari] field' (Islam, 2006, Par.15).

This well-organized movement, overwhelmingly supported by local communities and strengthened by the intellectual ideas and political support of the urban activist group NCBD, represents a successful case of subaltern resistance. Its mobilization strategies include both the politics of time and politics of scale, as mentioned earlier. Along with the Latin American cases analysed by Kirsch (2014), the Phulbari movement shows how the politics of time can work to an organization's advantage.

Dynamics of local mobilization

Although the anti-mining movement is widespread, grassroots protests are stronger in certain areas. This sub-section explains the historical and political constellation of three such areas, all of which have been involved in the Phulbari resistance. To maintain anonymity, these areas are called Areas 1, 2, and 3.

Area 1 is a historically significant place for peasant uprisings against British colonial power, and a number of active leftist organizations have inherited protest politics from the earlier peasant movement (e.g. *Tebhaga*) (Kamal, 2010). One contemporary example is the anti-mining movement; grassroots activists in this area consider foreign corporations to be plunderers of mineral resources around the world.[10] Area 2 consists of a cluster of villages where a non-local group called Chapaiyyas settled in the early 1970s. These people migrated from a distant region where they had lost homes and land due to river erosion. They colonized a large area of government-owned forest and cleared it for agriculture, but the majority have no land title.[11] The specific settlement history of this community has shaped their participation in the anti-mining movement in a unique way. Their relationship to the land is more complex than people living in other areas within the mining zone. The lack of legal documentation made them more vulnerable when faced with the resettlement and compensation package of the mining company.[12] They would not receive any financial compensation as they are not legally entitled to claim it.[13] The anti-mining movement in this area was a struggle to save a group of people who had worked very hard to build a community. Finally, Area 3 is a cluster of villages inhabited by half a dozen groups of *adivasis* (indigenous peoples). Historically, they have had a very contentious relationship with neighbouring Bengali Muslims who have taken much of their land (Soren et al., 2014). Forcible dispossession of property happens regularly in Bangladesh, and the state is indifferent to the plight of marginalized communities.[14] The Bangladeshi government, community leaders argued, had failed to protect them.[15] The anti-mining movement was an outburst against discrimination and marginalization. The coal mine, according to activists in this area, would destroy their unique community, culture, and heritage.

The voices of the people, their memories and histories of land, culture, and communities had no place in the narrative of the state and the mining company portraying the coal mine as necessary for economic growth. Not surprisingly, although activists in all three areas challenged this narrative, they did not form a homogenous group. There were variations in the lived experience of each community, making the concerted effort at resistance astonishing. Although local communities united against the mining company, the collectivity consisted of communities with antagonistic relationships.

For example, Bengali Muslims and *adivasis* (indigenous people) have quite different interests. One interviewee shared a story about the experience of mobilizing *adivasis* against the mining company. During an initial conversation, an *adivasi* community leader expressed some concerns about the nature of the movement. Most of the leaders of the anti-mining movement were Bengali Muslims. He feared that if there was any violence, Bengali Muslims leaders might push *adivasis* to the forefront and then not take part themselves, leaving *adivasis* to suffer more than Bengali Muslims.[16] This anecdote suggests the ongoing mistrust between these two communities. To explain this, we need to look at their history. Various groups of *adivasis* live in the northwestern region of Bangladesh. Documents of the British colonial administration (see Strong, 1912) show they have been living with Bengali Muslims and Hindus for a long period, but since the Partition of India in 1947, they have been discriminated against and exploited by Bengali Muslim communities (Knight, 2014). They have faced land dispossession and other forms of discrimination, leaving them marginalized and vulnerable. Kamal et al. (2006) call them *nijbhume parabasi* (alien in their own land). Land conflict is the dominant issue of the relationship between Bengali Muslims and *adivasis* in the region (Soren et al., 2014; Barkat et al., 2009; Kamal et al., 2006).

Similarly, the migrant people locally known as Chapaiyyas are stigmatized socially and culturally; they do not have harmonious relationships with their neighbouring communities even though they belong to the same Bengali Muslim group. Chapaiyyas represent a stigmatized social category, not well-accepted by other groups who consider them to have lower status. They identify Chapaiyyas as refugees.[17] Chapaiyyas are seen as close-knit communities which are different from their neighbours. Chapaiyyas used this cultural construction to strengthen their struggle against the mining company. They said the local communities had not accepted them as neighbours after more than 40 years of building community relations.[18] Where would they go if they were evicted?[19]

Anthropologist Fabiana Li (2015) offers a useful conceptual apparatus to understand solidarity among disparate groups at the grassroots level. She draws on Fortun's (2001) conceptualization of an 'enunciatory' community. According to Li, 'Enunciatory communities do not pre-exist, and they are not unchanging or internally coherent; they are not made up of members who share a common identity, but rather, they produce new identities' (2015, p. 6). This certainly applies to the Phulbari movement and the development of solidarity, despite ongoing tensions and differences in basic goals.

Post-2006 developments: contested coal rush

Although the Bangladeshi government refused to allow the mining company to proceed with its Phulbari plans after the violence in August 2006, the mining company did not give up. The Prime Minister instructed the Energy Ministry to cancel the Government of Bangladesh (GOB)'s contract with AEC,[20] but the mining company continued to lobby the government for approval of its mine development scheme.[21] Moreover, the development agency of the UK government (DFID) and the ambassador of the US government in Dhaka lobbied the Bangladeshi government on behalf of the mining company (Doward & Haider, 2006; Karim, 2010; Muhammad, 2014a). Based on the recommendation of a technical advisory committee (discussed shortly), the Prime Minister, who also held the portfolio of the Energy Ministry, initiated a plan in April 2010 to build an open pit coal mine at Barapukuria where a state-owned company already operated an underground mine (Daily Star, 2010).[22] The government was so persuaded by this idea that its Sixth Five Year Plan (2011–2015) planned to '[build up] mass awareness regarding the extraction procedure of coal especially for the open extraction method' (GOB, 2010a, p. 154). The Barapukuria and Phulbari coal mines are located in the same area in Dinajpur district.

The Prime Minister had ruled out the possibility of any open pit coal mine in Phulbari, but the parliamentary committee of the Energy Ministry strongly recommended constructing open pit coal mines in Phulbari and Barapukuria to meet the growing demand for coal. The committee visited German coal mines and, based on what it saw, suggested there was no risk in open pit coal mining. It urged the government to make a quick decision because the international community might ban coal mining in Bangladesh, bowing to the growing pressure of environmentalists around the world (Karmakar, 2011).

The government also unveiled a 20-year power system master plan (2010–2030) to generate electricity for the growing needs of the economy (GOB, 2011). The plan stipulated that 50 per cent of electricity would be generated by domestic primary energy, with domestic coal the prime source (25 per cent) (2011, p. 49). The mining company saw an opportunity to revive its agenda. It changed the narrative of the earlier export-oriented investment plan and framed the Phulbari coal mine as 'the source of energy for the people and business of Bangladesh' (GCM, 2009). It capitalized on the perennial electricity crisis and launched an aggressive campaign to influence policy makers. Its new frame argued that the coal mine would supply primary energy (i.e. coal) to generate 4000 MW of electricity for about 50 years. The mining company also submitted a new proposal to the Energy Ministry, suggesting that it would give the government a 10 per cent 'equity share' in addition to 6 per cent royalty (Karmakar, 2012). Business elites, corporate media, neoliberal think tanks, and local ruling elites largely supported this new agenda (BBC, 2012). NCBD and its local partners continued to campaign against any policy to build open pit coal mines in Bangladesh (NCBD, 2011). They organized a national convention in Dhaka in November 2009 and held a long march from Dhaka to Phulbari in October 2010. During 2012–2014, local activists resisted several initiatives of the mining company to begin its activities in Phulbari.

Considering the significance of these renewed protests, the Prime Minister announced that the Energy Ministry would wait for 'new technology' to extract coal without harming land, livelihood, and the environment (bdnews24, 2014). The PM also announced that the government would keep coal resources for future generations (Prothom Alo, 2012). Following this new policy directive, the Energy Ministry conducted several studies on hydrological issues in Barapukuria and on the possible effects of open pit mines in Barapukuria and Phulbari. These studies rejected the claims of the mining company (EMRD, 2016). An expert committee also assessed the feasibility study of the mining company and concluded that the study failed to take into account several critical issues, such as the loss of agriculture and environmental damage (Kalerkantha, 2012). The junior minister of the Energy Ministry[23] summarized the position of the Bangladeshi government:

> Our government puts peasants and their land first; coal is our second priority. The Prime Minister instructed [the Energy Ministry] to avoid energy projects that would destroy prime agricultural land, which grows 2–3 crops annually ... We can hire contractor to operate coal mines, if we do not have expertise. We need expertise to manage underground water resources. A unique feature of our geological condition [in Phulbari] is that there is a huge layer of water above the coal seam. We must know what will happen if we dewater the area to extract coal. There is no other example of such a coal mine in the world. I have visited many coal mines; our geological conditions are different from those coal mines. This is our coal resources, a national company will extract it, and coal resources will be utilized for the country's benefit (*desher kayla, desher company, deshei byabohar korbo*).

The last sentence reiterates the three key demands (three Nos) of the Phulbari movement: no open pit mining, no involvement of foreign mining company, and no export of mineral resources.

Recently, the Energy Ministry decided to ban open pit mining in Bangladesh entirely to protect precious arable land and underground water, two critical resources for Bangladesh's food security (Jahangir, 2016). It also unveiled a new power sector master plan to produce up to 25 per cent of electricity by 2041 using imported coal.[24]

Much of the recent change in focus can be attributed to the Phulbari movement. Its activists articulated a new set of ideas about coal mining in Bangladesh. They deconstructed corporate discourses of the mining company, created a set of demands, and mobilized mass support for those demands. These ideas/demands resonated with grassroots communities, showing how ideas drive social movements. Organic intellectuals (local and national), in the Gramscian sense, who were immersed in progressive political activism but had solid knowledge of the local issues, played a critical role in creating those ideas. Social movements require several significant mobilization processes to attract followers and to persuade policy makers: a critical diagnosis of the context to discredit corporate science (i.e. AEC's inaccurate articulation of the environmental consequences of the Phulbari mine), the articulation of a set of ideas, strategic meaning-making (i.e. crafting culturally resonant mobilization frames), and the education of the masses in local areas and beyond. In this case, the power elites bowed to the pressure and rethought their policy choices.

National scale mobilization: contesting the making of national coal policy

Although neoliberal restructuring of Bangladesh's economy began in the late 1970s (Quadir, 2000), the military regimes in the 1980s institutionalized the process by implementing a series of structural adjustment policies (Sobhan, 1991). In the energy sector, this began with the 'Energy Sector Adjustment Credit' of the World Bank (World Bank, 2004). A significant part of the adjustment credit was the 'Petroleum Exploration Promotion Project' (PEPP). The objective of PEPP was to attract foreign energy companies to explore oil and gas in Bangladesh. The project failed to achieve its objective (World Bank, 1993) but created investment opportunities for global capital and minimized the role of the state in the energy sector. In addition, a group of experts was hired by the World Bank to prepare a model 'production sharing contract' for petroleum exploration and to recommend necessary changes in national policies and laws. This shaped the formulation of Bangladesh's National Petroleum Policy in 1993 and National Energy Policy in 1996, thus institutionalizing the neoliberal transformation of the energy sector. Other relevant policies, such as investment policy, industrial policy and tax policy, were revised to keep pace with the new energy policy regime. The main objective was to set up a neoliberal framework conducive for the increased role of foreign private capital and the reduced role of the state.

As part of this policy thrust, the Bangladeshi government began to create opportunities for global energy companies. In the coal sector, it awarded a contract to an Australian mining company, BHP[25] to explore and develop coal resources in the northern region of the country. BHP was given many concessions, triggering criticism of the country's coal policy. In fact, debate over the neoliberal restructuring of the energy sector, particularly domestic coal development, originated in Phulbari when a local political activist questioned some of those concessions, saying they violated existing mineral law, and the deal with BHP failed to protect Bangladesh's national interests (see Bablu, 2005). Contentious issues included the meagre royalty and the right of the company to export a major portion of the resource without having to pay taxes (Rahman, 2008).

Against this backdrop, the Energy Ministry prepared a national coal policy to create an institutional framework for coal development projects. The Government of Bangladesh (GOB) recognized that the royalty rate and other issues needed revision, and a coal policy was required to regulate multinational mining corporations interested in developing domestic coal fields. GOB announced that the Phulbari coal mine would be approved *only* after the finalization of national coal policy. But this did not happen immediately; the making of national coal policy between 2005 and 2016 became a battleground at the national level between those endorsing neoliberal doctrines (e.g. GOB, AEC) and those wanting to keep state control over resource extraction (e.g. NCBD and its allies).

NCBD was formed in 1998 to mobilize popular protest against the privatization of resource extraction (Muhammad, 2004; Shaheedullah, 2002, 2009). Its members included leftist political parties, cultural activists, energy experts, academics, and intellectuals. NCBD's manifesto (*ghoshona*) reads:

> The primary goal of the National Committee is to establish people's control over [mineral] resources, in other words, to protect national resources from being grabbed by imperialist and hegemonic states and multinational corporations and to ensure the best use of these resources to serve the interests of the people. To attain this goal, the National Committee strives to cancel all anti-people contracts on national resources, including oil, gas, coal, and other mineral resources, as well as electrical power and seaports. It seeks to play a useful role in ensuring pro-people utilization of these resources and institutions. It also provides assistance to all political, social, cultural, and economic movements aligned with this goal.
>
> (NCBD, 2009, p. 1)

During the last two decades, it has organized several successful campaigns against multinational energy companies (see Faruque, 2017). After its strategic alliance with the local community movement in Phulbari,[26] NCBD spent much of its energy and resources confronting the Bangladeshi government and the mining company and deployed both discursive and direct-action tactics. Although it worked to strengthen local scale activism against the Phulbari coal mine, NCBD made coal policy its key issue.

Several independent experts and other civil society groups[27] supported NCBD's policy positions vis-à-vis coal development.

An initial draft of the government's coal policy was prepared in December 2005 by a state-owned consulting firm, supported by the World Bank and other Western development agencies (see Faruque, 2016). It recommended that investors should decide the mining method to use after assessing the economic viability and competitive energy markets. The policy allowed the exportation of coal with low royalty fees. It also recommended that two foreign mining companies be allowed to build open pit mines in Phulbari and Barapukuria (GOB, 2005). NCBD and various energy experts argued that the policy was biased towards foreign corporate interests. They suggested it did not consider the increasing domestic demand for coal resources (Kamal, 2006; Islam, 2010a). After several rounds of workshops with various stakeholders, such as government agencies, academics, civil society organizations, coal industry investors, and business associations, the Energy Ministry revised the policy and sent it to the Prime Minister's Office (PMO) for approval. The PMO sent it to an independent energy expert for his opinion. Based on the review report and the directives of the PMO, the Energy Ministry revised it again (Islam, 2010a).

The revised policy recommended higher royalty fees and stricter environmental guidelines. A low royalty rate would be applicable if the mining company sold coal *only* in the domestic market. If the mining company wanted to export coal, the royalty rate would be increased and the rate would be determined by a formula wherein the export value of coal was the main determining factor. In other words, instead of a fixed royalty rate, the policy proposed a variable one. It also recommended that about 40 per cent of the coal be reserved for domestic use (Khan, 2006a). Not surprisingly, the mining company expressed dissatisfaction with the revised policy suggesting that the policy was not driven by market forces. It would discourage foreign investment in coal mining, it said, and the export restriction would threaten the economic viability of mining (bdnews24, 2006; Khan, 2006b).

The Energy Ministry revised the policy to address these concerns and sent it to the PMO, then run by the Chief Advisor of the military-led Caretaker Government (CG).[28] The CG formed a technical advisory committee headed by an energy expert in June 2007 to review the policy and recommend appropriate changes. The committee consulted with various stakeholders, including local communities in the northern region,[29] and submitted its report in December (GOB, 2007).

NCBD submitted a memorandum to the technical advisory committee and demanded that the coal policy prohibit open pit mining and the exportation of mineral resources. They opposed any policy that would encourage ownership by foreign mining corporations. NCBD emphasized that Bangladesh was heavily dependent on natural gas, a supply likely to end in the near future; at that point, the only available primary energy would be coal. For NCBD, coal resources should not be left for foreign corporations to extract excessive amounts to increase their profits (NCBD, 2007). The technical advisory committee recommended establishing a public-sector coal development company, Coal Bangla, which would undertake coal exploration and development projects and engage foreign mining corporations, if necessary, through a joint venture project. It also suggested that exporting coal could be considered after ensuring Bangladesh's energy security for 50 years. Instead of a fixed royalty rate, it recommended that a 'Coal Sector Development Committee' should fix the royalty rate based on a number of factors, such as the cost of coal production, the price of coal in the international market, and the unit cost of coal-fired electricity. It also suggested that a small-scale open pit mine could be constructed in the northern part of Barapukuria to study the hydrological and environmental effects of such mining. If the results were satisfactory, the government could allow open pit mining on a commercial basis (GOB, 2007).

The Energy Ministry removed several critical issues from the report and submitted a revised policy for CG's approval (Islam, 2010a). NCBD and a group of public university teachers and eminent citizens urged CG not to approve the policy and to wait for a democratically elected government. Ultimately, CG did not make a decision and put it on hold. The new government invited a panel of non-resident Bangladeshi energy experts in June 2009 to examine the policy and suggest appropriate mining options. The panel recommended a goal of long-term energy security for the country, the strengthening of legal and institutional frameworks to address various environmental issues, and the creation of an environmental watchdog to ensure the transparency and accountability of mining corporations. They advised that the government not favour any particular mining method before adequately assessing geological, environmental, and socio-economic aspects (Khalequzzaman, 2010). Based on these recommendations, the Energy Ministry revised the policy in October 2010 and invited public opinion (GOB, 2010b).

NCBD provided an extensive critique of the proposed policy and offered alternative policy choices based on the three key demands mentioned previously – no open pit mining, no foreign mining corporations, and no export of mineral resources (three NOs). NCBD rejected the policy because, they argued, it favoured corporate interests and neglected public ones. It underscored that the memorandum signed between the government and the protesters in August 2006 should be the benchmark for national coal policy (NCBD, 2010). Meanwhile, independent experts suggested that many provisions in the proposed policy (e.g. coal production rate and coal export) were included to favour a specific mining company (Islam, 2010b).

Although the Sixth Five Year Plan said the policy would be finalized by the end of the planning period (GOB, 2010), during 2011–2015 there was no significant development. The Energy Ministry prepared a new coal policy in 2016 (Rasel, 2016a, 2016b). Several options partially address the demands of NCBD. For one thing, the government would prohibit export of mineral resources unless it was necessary to protect the public interest. For another, instead of allowing sole foreign ownership, the government would permit foreign mining corporations to start up joint ventures with local corporations. The policy proposed a production sharing contract (PSC) to attract foreign investment in mineral development. Finally, the government would keep a balance between coal demand and coal production to ensure long-term energy security.

This policy remains at the proposal stage, however. For the past ten years, the conflict between corporate and public interests has been so contentious that the Energy Ministry (regardless of the government in power) has failed to finalize a policy – and the conflict is ongoing. On the one hand, the mining company and its supporters continue to resist any progressive change in the coal policy because it will affect its profit margin. The mining company has informed potential global investors that it has an arrangement with the Bangladeshi government to keep the low royalty rate for the whole mining period, around 40 years (AEC, 2004). On the other hand, the activism of NCBD and its supporters continues to put enormous pressure on the Bangladeshi government to discard its neoliberal policy agenda in mineral development. As the approval of the Phulbari mine development scheme is contingent on the parameters set in the national coal policy, NCBD and its allies will try to derail it if it does not include their preferred items. An NCBD activist makes this clear:

> It will be impossible for the [Bangladeshi] government to accept [the demands of] NCBD. Because the neoliberal policy agenda, which it has been implementing [since the 1980s], is tied to the interests of its global development partners. The involvement of private sector in power generation, privatization of coal mines and gas blocks – all these are part of the same policy agenda. If the government wants to engage NCBD's agenda in its policy making process, this whole setup [the consensus of local and global elites] will not function. Our engagement will radically transform this process and the [Bangladeshi] government will not allow it. Therefore, a radical political struggle is required to change this process.

Conclusion

As this anti-mining movement shows, subaltern communities can displace the hegemonic discourses of extractive capital and the state. In Phulbari, the grassroots communities received tremendous intellectual and political support from urban activists whose anti-capitalist and anti-imperialist worldview (see Muhammad, 2004) helped transform their political consciousness. Urban activists did not ignore local issues to advance their own political agenda. Rather, they built on them and galvanized mass support by bridging local issues, as analysed in section one (i.e. land, livelihood, environment), and national questions, as presented in section two (i.e. state-capital consensus in the periphery, energy sovereignty). Careful articulation of culturally-resonant mobilization frames helped garner wide community support.

At the beginning of the Phulbari movement, bureaucratic and political elites were influenced by the corporate discourses of the mining company. Short-term economic benefits dominated the policy debate, with corporate media playing a significant role. But attention gradually shifted towards the perspectives of the challengers. The collaboration between local activists and NCBD and mass participation in local protest actions contributed to this shift. The activists clearly and collectively established the fact that social and environmental risks outweighed economic gains. Several GOB-commissioned studies (in 2006, 2012, 2014, and 2016) overwhelmingly supported their claims, and many independent experts and civil society groups offered robust critiques of the mining company. These efforts persuaded the Bangladeshi government to rethink its policy choices.

A key lesson for anti-mining movements is that mining companies deploy 'corporate science' to advance their economic agendas. NCBD and others identified significant flaws in 'scientific' studies of the mining company, specifically their underestimation of the social, economic, environmental, and political risks associated with coal mining in a densely populated agriculture-dependent community. Confronting corporate science, which enjoys the uncritical support of international financial institutions and Western development agencies,[30] powerful agents of neoliberal development in the global South, is not easy. It requires technical skills, organizational resources, and strong political support. Many local activists do not have these skills and resources. Members of the Phulbari movement recognized this weakness at the start of the movement and sought help from NCBD. Ultimately, the mining company and its governmental allies could not recover from the cumulative discursive assaults launched by NCBD and its allies.

The Phulbari movement offers another significant lesson for other anti-mining movements in the global South: those who use the 'politics of time' and 'politics of scale' can succeed. With respect to the former, the movement emerged as the mining company began pre-mining activities. The anticipatory politics of the challengers, based on credible scientific analysis of the context, created enormous pressure on the Bangladeshi government and forced it to retreat. As for the latter, scaling up the movement from a local to a national level increased the movement's tangible and intangible strategic resources.

The organizational form is a final point to consider. At the local level, the Phulbari movement was organized by community leaders well-known for their social and political activism, while at the national level, a radical activist group controlled the messages and resources and confronted powerful opponents. A shared goal brought local and national activists together, notably binding the quite diverse local groups in a common 'enunciatory' community (Li, 2015, p. 6). In the end, the mining company could not meet the challenge and the government yielded to the pressure.

Acknowledgement

This research was supported by the Munk School of Global Affairs, School of Graduate Studies; and the Department of Sociology at the University of Toronto. An earlier version of this chapter was presented at XIV World Congress of Rural Sociology at Ryerson University, Toronto (10–14 August 2016).

Notes

1 Interview with the Advisor to the Energy Ministry during his visit to Phulbari in July 2005. Source: AEC's documentary titled, Phulbari: Coal Capital of Bangladesh, available on YouTube. Retrieved 10 January 2015 from (www.youtube.com/watch?v=T-d2LzK9X7Q, accessed 10 January 2015).
2 Interviews in Phulbari, 2009.
3 Interview with leaders of Dinajpur Chamber of Commerce and Industry (DCCI), 2015.
4 Interview with leaders of *adivasi* (indigenous) communities in Birampur 2012.
5 CPP, Petition to the Prime Minister, 18 June 2005; Phulbari Community Council, Memorandum to Asia Energy (Bangladesh) Pty Ltd, 24 April 2005; Ganafront pamphlet, *Phulbari kaylakhani: kar ksati, kar labh*, July 2005.
6 Interviews with activists and informal group discussions with various local communities in Phulbari 2009, 2012.
7 Interview with a community leader in Phulbari, 2009.
8 Interview with the Mayor of Phulbari municipality (also a leader of CPP) in 2006 conducted by a Japanese NGO, Japan Center for a Sustainable Environment and Society (JACSES).
9 *Hartal* (a strike programme) closes everything down with the object of realizing a demand. The politics of *hartal* plays decisive role in mobilizing people and is a frequently used political tool for agitation.
10 Interviews with several activists of these political groups, 2009.
11 In-depth interview with a community leader, who has vast knowledge about the settlement history of this community, 2009.
12 My empirical findings are supported by historical studies on this community. For example, Cambridge historian Joya Chatterji (2013, p. 289) contends that Chapaiyyas are specialized in colonizing rich tracts of land and transforming them for farming. One of her informants said,

> We got this land only after we cleared this land and settled here ... We were living in Chapai [Nawabganj] and losing our land to the river; then one of

us got word that this place was a forest and that if we reclaimed it, it would belong to us.

13 Interviews and informal group discussions with local people, 2009, 2012.
14 Interview with a leader of a human rights organization, working for indigenous peoples living in northern districts in Bangladesh, 2013.
15 Interviews with two elderly community leaders, 2012.
16 Interview in Phulbari, 2012.
17 Interview in Birampur, 2009, 2012.
18 Informal group discussion in Birampur, 2012.
19 Interview in Birampur, 2012.
20 See 'Energy Division asked to move against Asia Energy Deal', *New Age*, 30 September 2006, p. 1. However, as of this writing, it remains pending.
21 See 'Asia Energy Lobbying for Phulbari coalfield', *New Age*, 28 February, 2007, p. 12.
22 The technical advisory committee formed by the Caretaker Government revived this idea. However, the technical debate about the prospect of open pit mining in northern Bangladesh was resolved in the early 1990s. A British mining consultant recommended underground mining for Barapukuria and the Bangladeshi government accepted (see Imam, 2013).
23 Talk Show on Independent TV, *Ajker Bangladesh*, 14 June 2015.
24 Ministry of Power, Energy and Mineral Resources, *Power System Master Plan 2016*, September 2016.
25 BHP discovered the Phulbari coal mine in 1997 and decided not to develop it. Two senior managers of BHP created a new company (AEC) and acquired the project in 1998. The Bangladeshi government approved the deal between BHP and AEC. See Faruque (2016) for an analysis of this deal.
26 CPP and NCBD signed a memorandum of understanding in August 2005.
27 Bangladesh Paribesh Andolon (BAPA), Bangladesh Environment Network (BEN), Citizen Commission of Bangladesh Economic Association, and Bangladesh *Adivasi* Forum.
28 Until the 15th amendment of Bangladesh's Constitution in 2011, there was a peculiar system of interim 'caretaker government' to oversee parliamentary elections. The tenure of this interim government was three months. There was political chaos in 2006; main political parties were in a confrontational mood over the formation of the interim government. The interim government which took over in late 2006 was ousted by a military-led caretaker government in January 2007; the latter remained in power until the end of 2008.
29 All coal fields discovered so far are in this region.
30 International financial institutions also ignore these critical issues; for example, the feasibility study of the mining company, which was widely discredited, was approved by the Asian Development Bank (ADB), the political risk insurance underwriter of the mining company.

References

AEC. (2004). *The Phulbari Coal Project: Presentation to Investment Professionals*. London: Asia Energy Plc (AEC).
AEC. (2005). *Phulbari Coal Project: Bulletin of Asia Energy (October 2005)*. Phulbari: Asia Energy Corporation (Bangladesh) Pty Ltd (AEC).
AEC. (2006). *Phulbari Coal Project: Environmental and Social Impact Assessment (Volume 1)*. Dhaka: Asia Energy Corporation (Bangladesh) Pty Ltd (AEC).
Ahmed, N. (2013). Entangled earth. *Third Text*, 27, 1, 44–53.

Bablu, A. I. (2005). Phulbari kayla sampad raksay egiye ashun: Samrajyabad o kamishanbhogi sashak-gosthir hat theke jatiya sampad raksa karun, *Anirban*, June-August, 9–11.

Barkat, A., Hoque, M., Halim, S., & Osman, A. (2009). *Life and Land of Adibashis: Land Dispossession and Alienation of Adibashis in the Plain Districts of Bangladesh*. Dhaka: Pathak Shamabesh, Dhaka.

BBC. (2012). Bangladesh Sanglap, *BBC Bangla*, 8 December. Retrieved 20 December 2012 from www.youtube.com/watch?v=rxXvDQw6Jo0.

BBS – Bangladesh Bureau of Statistics. (2014). *Bangladesh Population and Housing Census 2011: Community Report, Dinajpur Zila*. Dhaka: Bangladesh Bureau of Statistics.

Bebbington, A. (Ed.). (2012). *Social Conflict, Economic Development and Extractive Industry: Evidence from South America*. London: Routledge.

Bebbington, A. & Burry, J. (Eds). (2013). *Subterranean Struggles: New Dynamics of Mining, Oil, and Gas in Latin America*. Austin, TX: University of Texas Press.

Bdnews24. (2006). Decisions on Phulbari coalmine development to be in line with agreement: Advisor, 7 February. Retrieved 10 January 2010 from https://bdnews24.com/business/2006/02/07/decision-on-phulbari-coalmine-development-to-be-in-line-with-agreement-advisor.

Bdnews24. (2014). Kayla: 'natun prajuktir' opeksa korbe sarkar, 6 February. Retrieved 7 February 2014 from https://bangla.bdnews24.com/bangladesh/article740267.bdnews.

Butler, P. (2015). *Colonial Extractions: Race and Canadian Mining in Contemporary Africa*. Toronto: University of Toronto Press.

Chatterji, J. (2013). Dispositions and destinations: Refugee agency and 'mobility capital' in the Bengal Diaspora, 1947–2007. *Comparative Studies in Society and History*, 55, 2, 273–304.

Chowdhury, N. S. (2016). Mines and signs: resource and political futures in Bangladesh. *Journal of the Royal Anthropological Institute*, 21, S1, 87–107.

Collier, P. (2010). *The Plundered Planet*. New York: Oxford University Press.

CPP – Committee to Protect Phulbari (2005a). Leaflet, June.

CPP – Committee to Protect Phulbari (2005b). Petition to the Prime Minister, 18 June.

DAE – Department of Agricultural Extension. (2017). *Zila Agricultural Statistics*. Dinajpur: Department of Agricultural Extension.

Daily Star. (2010). Barapukuria Coal Mine: PM talks option for open pit, 11 April. Retrieved 12 April 2010 from www.thedailystar.net/news-detail-133832.

Deonandan, K. & Dougherty, M. L. (Eds). (2016). *Mining in Latin America: Critical Approaches to the New Extraction*. London: Routledge.

Doward, J. & Haider, M. (2006). The mystery death, a town in uproar and a $1bn UK mines deal, *Guardian*, 3 September. Retrieved 15 January 2010 from www.theguardian.com/world/2006/sep/03/bangladesh.

EMRD – Energy and Mineral Resources Division. (2016). *Annual Report 2015–2016*, Dhaka: Energy and Mineral Resources Division.

Faruque, M. O. (2016). 'Phulbari khani ebang Bangladesher jwalani rajaniti'. *Sarbojonkotha*, 2, 4, 49–57.

Faruque, M. O. (2017). Neoliberal resource governance and counter-hegemonic social movement in Bangladesh. *Social Movement Studies*, 16, 2, 254–259.

Fortun, K. (2001). *Advocacy after Bhopal: Environmentalism, Disaster, New Global Orders*. Chicago, IL: University of Chicago Press.

Gardner, K. (2012). *Discordant Development: Global Capitalism and the Struggle for Connection in Bangladesh*. London: Pluto Press.

GCM. (2009). *Annual Reports and Accounts*. London: GCM Resources plc.

GCM. (2016). *Annual Report and Accounts*, London: GCM Resources plc.

Giugni, M. et al. (Eds). (1999). *How Social Movements Matter*. Minneapolis, MN: University of Minnesota Press.

GOB – Government of Bangladesh. (2005). *Bangladesh Coal Policy – Draft*. Dhaka: Energy and Mineral Resources Division, Government of Bangladesh.

GOB – Government of Bangladesh. (2007). *Report of the Bangladesh Coal Policy Advisory Committee*. Dhaka: Energy and Mineral Resources Division, Government of Bangladesh.

GOB – Government of Bangladesh. (2010a). *Sixth Five Year Plan: 2011–2015 (Part-2)*. Dhaka: Planning Commission, Government of Bangladesh.

GOB – Government of Bangladesh. (2010b). *Bangladesh kayla niti-2010 prastabita khasra*. Dhaka: Energy and Mineral Resources Division, Government of Bangladesh.

GOB – Government of Bangladesh. (2011). *Power System Master Plan 2010*, Dhaka: Ministry of Power, Energy and Mineral Resources.

GOB – Government of Bangladesh. (2015). *Seventh Five Year Plan: 2016–2020*. Dhaka: Planning Commission, Government of Bangladesh.

GOB – Government of Bangladesh. (2016). *Power System Master Plan 2016*, Dhaka: Ministry of Power, Energy and Mineral Resources.

Horowitz, L. S. & Watts, M. J. (Eds). (2017). *Grassroots Environmental Governance: Community Engagements with Industry*. London: Routledge.

Horowitz, L. S. (2015). Culturally articulated neoliberalisation: corporate social responsibility and the capture of indigenous legitimacy in New Caledonia. *Transactions of the Institute of British Geographers*, 40, 88–101.

Humphreys, M., Sachs, J., & Stiglitz, J. E. (Eds). (2007). *Escaping the Resource Curse*. New York: Columbia University Press.

Imam, B. (2013). *Energy Resources of Bangladesh* (2nd edition). Dhaka: University Grants Commission of Bangladesh.

Islam, A. (2006). Report on Asia Energy's dev Scheme submitted to govt. *New Age*, 25 September, p. 1.

Islam, M. N. (2010a). Kaylaniti: prothom theke dosh khashrar purbapor, *Prothom Alo*, 13 December. Retrieved 14 December 2010 from http://archive.prothom-alo.com/detail/date/2010-12-13/news/115410.

Islam, M. N. (2010b). Kaylaniti: dasham songskoroner khasra niye kichu motamot, *Prothom Alo*, 23 December. Retrieved 24 December 2010 from http://archive.prothom-alo.com/detail/date/2010-12-23/news/117717.

Jahangir, S. (2016). Govt backtracks on open-pit mining in Barapukuria, Phulbari, *Daily Sun*, 23 September. Retrieved 24 September 2016 from www.daily-sun.com/arcprint/details/169050/Govt-backtracks-on-openpit-mining-in-Barapukuria-Phulbari/2016-09-23.

Kalerkantha. (2012). Unkumta paddhotite kayla uttoloner prastab bisheshoggo Committee'r, 10 June. Retrieved 11 June 2012 from www.kalerkantho.com/print-edition/world/2012/06/10/260501.

Kamal, M., Chakrabarty, I., & Nasrin, Z. (2006). *Nijbhume Parabasi: Uttarbanger Adivasir Prantikota Discourse*. Dhaka: Dibyaprakash.

Kamal, N. (2006). *Prostabito koyla niti: ekti bishleshon*. Dhaka: Bangladesh Economic Association.

Karim, F. (2010). WikiLeaks cables: US pushed for reopening of Bangladesh coal mine. *Guardian*, 21 December. Retrieved 21 December 2010 from www.theguardian.com/world/2010/dec/21/wikileaks-cables-us-bangladesh-coal-mine.

Karmakar, A. (2012). Phulbari kayla khoni unnayane Asia Energy'er natun prastab, *Prothom Alo*, 4 November. Retrieved 5 November 2012 from http://archive.prothom-alo.com/detail/date/2012-11-04/news/302620.

Karmakar, A. (2011). Unmukta kaylakhanir pokse mot dilo sangsadiya pratinidhidal, *Prothom Alo*, 5 February. Retrieved 6 February 2011 from http://archive.prothom-alo.com/detail/news/129110.

Khalequzzaman, M. (2010). The Quarry quandary. *Daily Star*. 1 June. Retrieved 2 June 2010 from www.thedailystar.net/news-detail-140830.

Khan, S. (2006a). 'New coal policy eyes higher royalty, environment safety: Installation of power plant a must for mine developers', *Daily Star*, 9 February. Retrieved 15 February 2010 from http://archive.thedailystar.net/2006/02/09/d6020901033.htm.

Khan, S. (2006b). 'Coal Policy may get cabinet nod tomorrow: Export to yield higher royalty. *Daily Star*, 9 July. Retrieved 15 February 2010 from http://archive.thedailystar.net/2006/07/09/d6070901011.htm.

Kirsch, S. (2014). *Mining Capitalism: The Relationship between Corporations and Their Critics*. Oakland, CA: University of California Press.

Klandermans, B. (1997). *The Social Psychology of Protest*. Oxford: Blackwell Publishers.

Knight, F. (2014). *Law, Power and Culture: Supporting Change from Within*. New York: Palgrave Macmillan.

Kumar, K. (2014). The sacred mountain: Confronting global capital at Niyamgiri. *Geoforum*, 54, 196–206.

Le Billon, P. (2012). *Wars of Plunder: Conflicts, Profits and the Politics of Resources*. New York: Columbia University Press.

Li, F. (2015). *Unearthing Conflict: Corporate Mining, Activism, and Expertise in Peru*. Durham, NC: Duke University Press.

Luthfa, S. (2011). 'Everything changed after the 26th': Repression and Resilience against proposed Phulbari Coal Mine in Bangladesh. *QEH Working Paper Series*, 193, 1–22.

Macdonald, K., Marshall, S., & Balaton-Chrimes, S. (2017). Demanding rights in company-community resource extraction conflicts: Examining the cases of Vedanta and POSCO in Odisha, India. In J. Gruel, J. N. Singh, L. Fontana, & A. Uhlin (Eds), *Demanding Justice in the Global South: Claiming Rights* (pp. 43–67). New York: Palgrave Macmillan.

McMichael, P. (Ed.). (2010). *Contesting Development: Critical Struggles for Social Change*. New York: Routledge.

Moody, R. (2005). *The Risks We Run: Mining, Communities and Political Risk Insurance*. Utrecht: International Books.

Muhammad, A. (2003). Bangladesh's integration into global capitalist system: Policy direction and the role of global institutions. In M. Rahman (Ed.)

Globalisation, Environmental Crisis and Social Change in Bangladesh (pp. 113–140). Dhaka: University Press Limited.

Muhammad, A. (2004). Samaj rupantorer biplobi lorai ebong bamponthider nora-chora. *Natunpath*, 3, 1, 55–71.

Muhammad, A. (2006). Phulbari and the People's Verdict. *Daily Star*, 24 September, http://archive.thedailystar.net/2006/09/24/d609241501130.htm (accessed 15 January 2010).

Muhammad, A. (2013). Phulbari protirodh: natun unnayan prashna. *Bonik Barta*, 26 August, p. 4.

Muhammad, A. (2014a). Natural resources and energy security: Challenging the 'resource-curse' model in Bangladesh. *Economic and Political Weekly*, 49, 4, 59–67.

Muhammad, A. (2014b). Phulbari ganaabhyutthan: unnayaner malikana. *Bonik Barta*, 26 August, p. 4.

NCBD – National Committee to Protect Oil, Gas, Mineral Resource, Power and Ports. (2005). *Phulbari kayla prakalpa kar labh kar ksati*. Dhaka: NCBD.

NCBD – National Committee to Protect Oil, Gas, Mineral Resource, Power and Ports. (2007). *Memorandum to the Convenor of the Bangladesh Coal Policy Review Committee*. Dhaka: NCBD.

NCBD – National Committee to Protect Oil, Gas, Mineral Resource, Power and Ports. (2009). *Ghoshona o parichalona nitimala*. Dhaka: NCBD.

NCBD – National Committee to Protect Oil, Gas, Mineral Resource, Power and Ports. (2010). *Bangladesh kaylaniti 2010 (prastabita khasra) er upor jatiya committee'r mantabya o baktabya*. Dhaka: NCBD.

NCBD – National Committee to Protect Oil, Gas, Mineral Resource, Power and Ports. (2011). *Barapukuriya ba phulbari ba desher anya kothao unmukta pad-dhatite kayala khanan pratirodh karate habe keno?* Dhaka: National Committee to Protect Oil, Gas, Mineral Resource, Power and Ports.

Nuremowla, S. (2016). Land, place and resistance to displacement in Phulbari. *South Asia Multidisciplinary Academic Journal*, 13, 1–17.

Ozen, H. & Ozen, S. (2017). What makes locals protesters? A discursive analysis of two cases in gold-mining industry in Turkey. *World Development*, 90, 256–268.

Prothom, A. (2012). 'IEB'r sammelone prodhanmontri: Kayla bhabisyat prajanmer jonno mojud thakuk', 15 January. Retrieved 16 January 2012 from http://archive.prothom-alo.com/detail/date/2012-01-15/news/216631.

Quadir, F. (2000). The political economy of pro-market reforms in Bangladesh: Regime consolidation through economic liberalization? *Contemporary South Asia*, 9, 2, 197–212.

Rahman, M. (2008). *Phulbarir rajniti*. Dhaka: Kashbon Prakashan.

Rajak, D. (2011). *In Good Company: An Anatomy of Corporate Social Responsibility*. Stanford, CA: Stanford University Press.

Rasel, A. R. (2016a). Coal policy draft ready to be finalised. *Dhaka Tribune*, 4 April. Retrieved 5 April 2016 from www.dhakatribune.com/bangladesh/2016/apr/04/coal-policy-draft-ready-be-finalised.

Rasel, A. R. (2016b). Coal sector to get production sharing contract. *Dhaka Tribune*, 7 April. Retrieved 8 April 2016 from www.dhakatribune.com/bangladesh/2016/apr/07/coal-sector-get-production-sharing-contract.

Shaheedullah, S. M. (2002). Gyas-sampad lunthan o bidyut-bandar khetre agrasan er biruddhe pratirodh andolan, *Natun Diganta*, 1, 2, 219–238.

Shaheedullah, S. M. (2009). Jatiya khanija sampader upar bahujatik kompanir agrasan, *Halkhata*, 3, 4, 21–28.

Singh, J. T. N. & Camba, A. A. (2016). Neoliberalism, resource governance and the everyday politics of protests in the Philippines. In J. Elias and L. Rethel. (Eds) *The Everyday Political Economy of Southeast Asia* (pp. 49–71). New York: Cambridge University Press.

Sobhan, R. (Ed.) (1991). *Structural Adjustment Policies in the Third World: Design and Experience*. Dhaka: University Press Limited.

Soren, R., Borhan, A., & Sumon, M. (Eds). (2014). *Jarip Theke Bayan: Bangladesh Rashtrer Uttar-Pashchimanchaler Prantik Jatisattar Manusher Bhumisamasyakendrik Samikkha O Bayan*. Dhaka: Sangbed.

Strong, F. W. (1912). *Eastern Bengal District Gazetteers: Dinajpur*. Allahabad: The Pioneer Press.

Walker-Said, C. & Kelly, J. D. (Eds). (2015). *Corporate Social Responsibility? Human Rights in the New Global Economy*. Chicago, IL: University of Chicago Press.

World Bank. (1993). *Performance Audit Report: Petroleum Exploration Promotion Project (Report No. 12114)*. Washington, DC: World Bank.

World Bank. (2004). *Project Performance Reassessment Report: Energy Sector Adjustment Credit (Report No. 29379)*. Washington, DC: World Bank.

5 Local struggles for the coproduction of natural capital

Payment for forest environmental services in Central Vietnam

Fumikazu Ubukata and Truong Quang Hoang

Introduction: the creation of 'natural capital' and local villagers

Social movements are rarely successful in completely stopping natural resource development projects but they sometimes influence the impact on local communities most directly affected by a project. Compensation is a strategy governments and private sector proponents can use to reduce resistance and to some degree distribute project benefits. One approach to compensation has been through payments for environmental services.

The payment for environmental services (PES)[1] approach generates financial flows between providers and users of environmental services (Wunder, 2005). Like other market approaches, PES generally assumes a tradeoff between providers' livelihood activities and natural resource development (NRD). This contrasts greatly with 'community management' approaches, which assume positive linkages between rural livelihood and NRD (Wunder, 2005). With PES, nature is viewed as a valuable asset that can provide various environmental services and can benefit society and the economy in the long term. However, according to conventional environmental economics, environmental services are not internalized in our economic calculations. Thus, what PES can do is help to internalize the costs of environmental services by creating a link between environmental service flows and financial flows.

Commodification of nature is not new. Polanyi (1957) recognized it as a central process in the 'great transformation' to a market society. Commodification has expanded as our economy has become globalized, neo-liberalized, and financialized (Epstein, 2008; Levitt, 2013). Commoditization is the 'solution' provided by environmental economics for the decline of the natural world. By creating a financial value for 'natural capital' nature is incorporated into our economic activities (Daily & Ellison, 2002, Tercek & Adams, 2013).

The logic is very similar to that found in Hernando de Soto's argument regarding the activation of 'dead capital'. According to de Soto (2000), one of the main reasons that capitalism fails in many developing countries is not because people do not have enough capital, but because they neglect the potential of existing assets, which are 'dead' due to political informality. He argues that recognizing the potential of this 'dead capital' and activating it through formalization is one way to activate the economy of the poor and even upgrade their assets. In this sense PES regards nature as 'natural capital' and transforms 'dead' natural capital into 'active' natural capital. It is therefore possible to view PES as a new phase in the capitalization of nature, following the efforts of land capitalization through land titling (or allocation) programmes that transform land into a valuable economic asset.

Meanwhile, from a local perspective, PES is a way to mitigate resistance to NRD projects. For instance, some argue that PES can establish equitable (or at least 'fair') environmental markets through the 'provider gets principle', which holds that those who provide an environmental benefit should be rewarded for doing so (Pagiola, Bishop, & Landell-Mills, 2002: p. 6). Some also argue that PES is more cost-effective than 'community management' approaches, such as integrated conservation and development projects (ICDPs) (Ferraro & Simpson, 2002). How well PES actually works depends on how the market for environmental services is constructed and implemented. The process that creates and implements PES greatly affects the local response.

This chapter examines the processes and local responses to such a capitalization of nature programme called Payment for Forest Environmental Services (PFES) in Central Vietnam's Thua Thien Hue province. The information provided here is based on a series of interviews with relevant stakeholders (i.e. national-level policy makers, provincial officers, technical staff, local officers, and villagers) during our 2015 and 2016 field surveys. We first briefly introduce the PFES programme in Vietnam. Second, we describe how it has been implemented in Thua Thien Hue province. Third, we describe the local response in a village resettled due to dam construction, one of the cases most affected by the exploitation of 'forest environmental services' (FES) in the area. We conclude with a discussion of how different meanings of nature are practiced and negotiated in this authoritarian sociopolitical setting.

PFES policy and its implementation in Central Vietnam

PFES as the Vietnamese version of PES

Vietnam is the first country in Asia to institutionalize PES as a nationwide policy (Pham, Bennett, Vu, Brunner, Le, & Nguyen, 2013). The Payments for Forest Environmental Services (PFES) is 'a service payment mechanism between FES [forest environmental service] users and providers, with an aim to mobilize contributions from society for forest protection and to enhance the economic value of the forest environment' (VFDP, n.d., 8).

The programme was piloted in 2008 in Lam Dong and Son La provinces and was expanded nationwide in 2010 by the enactment of Decree 99/2010/ND-CP (Decree 99 hereafter). By 2014, 162 million USD had been paid to environmental service providers through this programme (VFDP, n.d.).

In this programme, FES is defined as 'the provision of use values from the forest environment to meet the needs of the society and people's life' (VFDP, n.d., 12). In Decree 99, five types of services are officially included, and three types of service users are recognized (Table 5.1).

Of these, hydropower plants are currently the dominant user category and occupy more than 65 per cent of the total payment contracts in the programme. Meanwhile, service providers that are entitled to payments are categorized as follows: a) forest owners as organizations, such as

Table 5.1 Types of defined FES and the actual users

FES type	Main FES users	Service charge	Number of contracts (as of Aug. 2014)
1) Soil protection, reduction of erosion and sedimentation of reservoirs, rivers, and streams	Hydropower plants	20 VND/kwh of commercial electricity	235 (66.9%)
2) Regulation and maintenance of water sources for production and social life	Water supply companies	40 VND/m³ of commercial water	72 (20.5%)
3) Forest carbon sequestration and retention, reduction of greenhouse gas emissions by preventing forest degradation and forest area decline, and developing forests in a sustainable manner	None	–	–
4) Protection of natural landscape and conservation of ecosystem biodiversity for tourism	Tourism agencies	1–2% of total revenue	44 (12.6%)
5) Provision of spawning grounds, sources of feed and natural seeds, and use of water from forest for aquaculture	None	–	–

Source: VFDP (n.d.).

Forest Management Boards (FMBs)[2] and forest enterprises; b) forest owners who are households, individuals, or residential communities; and c) households, individuals, and residential communities who are contracted for forest protection with state forest owners (VFDP, n.d., p. 19). The PFES programme connects these FES users and providers with financial flows as shown in Figure 5.1. Through these payments, PFES aims to achieve forest conservation in upstream watersheds and improve villagers' livelihoods (Pham et al., 2013).

To implement PFES, it is necessary to construct a technical and institutional foundation by identifying service users and providers, evaluate the forest as a source of services, and determine payment rates. For this purpose, a Forest Protection and Development Fund (FPDF) was established for each province containing a forest. A national level fund was also established in 2008 under the management of the Ministry of Agriculture and Rural Development (MARD). This is the Vietnam Forest Protection and Development Fund (VNFF).

The FPDF also acts as an intermediary between service users and providers, as it collects payments from users and distributes payments to service providers (Figure 5.1). The actual institutionalization processes largely depend on provincial-level FPDFs, which reflect provincial social and administrative settings and have led to different levels of implementation among the provinces. A fixed rate (see Table 5.1) is applied for the collection of users' fees. Payments to service providers are determined at the watershed level. The total fees collected from users in a watershed are then divided into the area under forest vegetation adjusted by the K coefficient.[3]

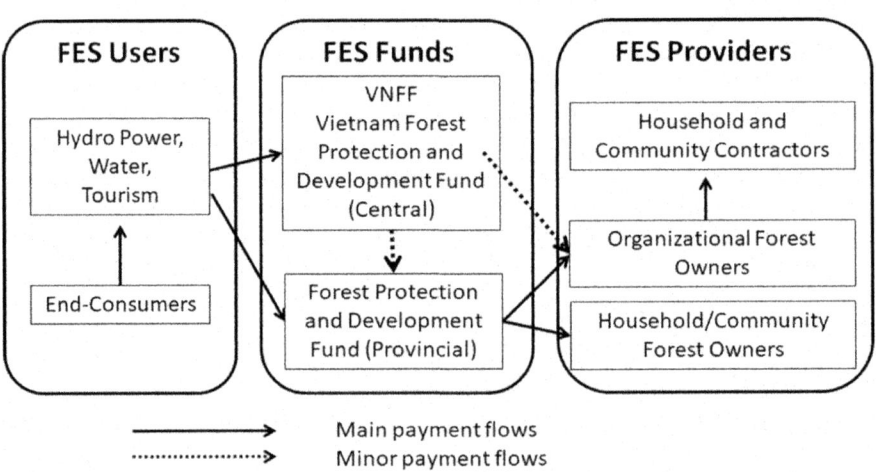

Figure 5.1 Vietnam payment for environmental services scheme in brief.

Source: VFDP (n.d.) and field survey. Figure by authors.

Since different watersheds vary in size, have different users, and have differing numbers of service providers, this system produces remarkable differences in payments received by service providers, thus causing unequal payments among watersheds (VFDP n.d.).

Implementation process of the PFES in Thua Thien Hue province

How are technical and institutional conditions produced in the PFES implementation process? In Thua Thien Hue province, the process has three steps. First, service providers and their managing territories are identified and mapped and the watershed area for payment is identified. FPDF staff and delegated consultants use GIS techniques to integrate various digital maps, including forest classification maps and forest vegetation maps, along with forest owner information. Finally, they create an integrated map containing all of this information.

Second, the map is verified at the commune or forest management board (FMB) and at village levels and is revised if necessary. This process eventually leads to a short field visit by FPDF staff or forest rangers. Based on the map, conditions for individual contracts are prepared. At this stage, the payment rate information is not yet available, but villagers must decide whether to agree to the conditions. If they agree (and in most cases they do), they sign a contract.

Third, based on these contracts, the payment rate is ultimately decided and then the providers (villagers and FMBs) are informed. The rate is set by the Provincial People's Committee (PPC). This means the providers have to sign the contracts without knowing the final payment rate.[4]

The overall implementation of PFES in Thua Thien Hue province was delayed. Providers did not begin receiving payments until 2014 even though the provincial FPDF was established in 2011, just after the enactment of Decree 99. There are two reasons for this delay. First, the provincial FPDF was short on funding and human resources; only $530,000 US had been collected from service users during 2011–2013, ranking twenty-first among all PFES provinces in Vietnam (VFDP n.d.). By 2015, only three hydropower plants were engaged in the programme, which limited the amount of contributions. There were only 12 staff members in the FPDF (as of 2015), including four who also worked in the provincial forest office.

The second reason for delay was the complex and fragmented 'forest ownership' in the province. As a former socialist nation, the land in Vietnam legally belongs to the state. The Vietnamese government began issuing long-term usufruct rights to the forest through the forest allocation programme in the 1990s. Recipients of such rights ranged from large holders (e.g. FMBs and forest enterprises) to smallholders (e.g. groups of villagers and individual households). In Thua Thien Hue province, individual households are the dominant recipients of such rights. According to the FPDF officers, over 500 qualified 'forest owners' were identified in the province by 2015. This number contrasts with that of neighbouring Quang Nam province, which had only 15 qualified 'forest owners' (mostly FMBs). This fragmented ownership structure has made the transaction costs of implementation very high.

PFES as the capitalization of nature

The PFES policy and implementation process can be interpreted as a part of a three-stage strategic process of capitalization as described by de Soto (2000): 1) the discovery strategy (the process in which policy makers 'discover' specific resources to capitalize), 2) the political and legal strategy (the process in which policy makers remove political and legal obstacles for capitalization), and 3) the operational strategy (the actual operational process for implementing the project). It is crucially important, therefore, to consider how and by whom these three strategies were created.

Regarding the discovery strategy, policy makers' ideas significantly affected the definition of the FES and the programme's overall structure. One of the programme's founders stated in an interview that he and his colleague modified the PES concept to fit 'Vietnamese circumstances', whereby they revised its objective from maintaining ecosystem services to conservation of forests. As will be shown later, they also discussed FES as a way to ration payments to forests, and finally introduced the idea that forest environments have use value created by people's labour in forest management.

These policy makers' discussions were then reflected in the political and legal strategy, such as in the official definition of FES and the programme's structure in Decree 99. As shown in Table 5.1, several assumed functions of forests were selectively defined as environmental services, of which water provision to hydropower dams in the watershed was regarded as the most important. This means more FES were recognized when more hydropower dams were constructed.

Finally, producing the technical and institutional conditions at the local and provincial levels is of crucial importance in the operational strategy. As a state funding agency, FPDF plays a significant role in creating these bases and mediates financial flows. With the FPDF's direction information is centralized and compiled at the provincial level using information technology (e.g. GIS), which creates a database to identify, evaluate, and decide who pays, for which forest they pay, and how much should be paid into the programme.

In short, all of the information described here indicates that FES is politically and economically 'founded' and legislatively 'created', as it is based on an official idea influenced by the concept of environmental services. The following two dimensions of 'politico-economic constructions' are particularly important when we consider the impact on local responses.

First, it is apparent that the state is dominant in all strategies. In that sense, the process does not follow 'orthodox 'neoliberalization of nature' approaches in their use of market instruments' (McElwee, 2012, p. 412). Instead, the PFES programme can almost be regarded as a form of taxation rather than a private-market mechanism. However, PFES policy differs from taxation in that it does try to connect FES users and providers. A reasonable and justifiable flow between environmental services and financial payments is required. State involvement delayed the process as a result of complex local rights structures and limited available resources. In actual implementation local social contexts influenced outcomes even though all processes were managed by the state.

Second, it is crucial that FES is founded and defined by a small number of people – namely national-level policy makers. As we will show later, FES can be interpreted differently by different stakeholders in the programme. Intentional or not, multiple 'meanings' of FES are accommodated within the institutional design. This can be a source of misunderstanding and conflict among stakeholders.

Villagers' responses on the brink of capitalization

Relocation of the BH village

We now turn to how villagers, as important PFES recipients, reacted to the programme, particularly in terms of the problems produced by technical and institutional conditions as well as the implementation process itself. Shortly, we provide a case study of BH village in the province. This is not a typical case in which villagers did receive payments, rather, this case demonstrates the contradictions inherent in the programme. This village is one of the most greatly affected by the exploitation of FES in the area. As such, it demonstrates how different ideas about nature were negotiated within an authoritarian political setting.

With a population of 252 villagers (57 households), BH village is a resettlement village established after the original village area became a hydropower dam site in 2006. As explained in the previous section, the construction of hydropower dams is what initiates PFES in many cases. Without dam construction, most FES are not officially recognized and thus are not 'created'. In this sense, the villagers in BH village were victims of relocation due to dam construction but were also turned into important stakeholders through FES creation.

Most villagers were part of the Co Tu ethnic minority, which had spontaneously migrated from the mountainous areas upstream to an earlier village site during the 1990s. As their relationship with the official administration was limited prior to relocation in 2006, they were able to maintain their traditional livelihoods, which included slash-and-burn agriculture combined with bamboo and fruit tree planting. They collected various forest products from the nearby forest, such as rattan, bee honey, herbs, and even timber. They also maintained their traditional social order where the head of the clan (the patriarch, called *Gia Lang*) governed his lineage in the village.

To them, the forest was a communal resource that was loosely managed by their traditional social order. Although their cultivated land is common property, it becomes de facto 'private' when perennial crops such as bamboo and trees are planted. Those who had additional labour time and money accumulated assets by planting these perennial crops. The pre-relocation area, however, was an officially protected forest; thus, it was officially illegal to live in that area and exploit its resources. Some years before the relocation, the provincial authority informed them about the hydropower dam construction downstream. The authorities and the company initiated a relocation scheme, and the villagers had no choice but to accept it.

Villagers' livelihoods

After the resettlement, the villagers' livelihoods changed completely. Many households received some compensation from the hydropower company for their houses and crops. However, they did not receive land compensation as the land belonged to a state-owned protection forest area. Villagers were provided with new houses and housing areas in the resettlement location. However, what was crucial for them was that they did not receive equivalent cultivated land. Instead they received 2,500 square metres of unproductive land. Hence, many villagers fell into poverty after the resettlement because of the lack of land for cultivation.

In response, the villagers combined several options according to their household conditions. First, those who could provide labour engaged in wage work such as construction, acacia harvesting, and processing. However, these job opportunities were limited, and sometimes the working conditions were harsh. Second, some were lucky enough to keep access to their original land if it was not flooded by the dam. Some had applied for the state-sponsored acacia planting project and had planted acacia in the past since the project had land allocation for planting as a precondition. A few other households were also still productive, as they were aware of the dam construction before the official announcement and had begun reclaiming land outside the pre-relocation area and they retained access to this land after relocation.

Third, some villagers tried to reclaim land near the original site. In most cases, they attempted planting bamboos and acacia trees. However, this was illegal as the area now belonged to the dam watershed and was classified as a protection forest area under the FMB's management. For the most part, villagers were aware that their activity was illegal but stated they had to cultivate the land for survival. A few unlucky households were ejected from their plots when FMB staff detected their forest encroachment. As of the late 2000s, the FMB's forest monitoring activities had strengthened and the introduction of the PFES programme increased the number of ejections. One FMB officer in the area confirmed that they employed more field staff to monitor the forest since it received PFES payments. BH villagers also observed that they had been expelled more frequently in recent years.

One villager complained that he could not understand why some people were able to continue cultivation in that area while others were not. He also suspected that those with personal connections to the staff were covertly allowed to cultivate the land, leading to an unfair situation. Further, it is unlikely that the FMB staff can detect all violations. The FMB officers themselves recognize this reality. As the area is far from the core FMB management zone, the area is very hard for the FMB to manage.

Thus, the situation was much like a cat-and-mouse game in which the villagers were practicing forms of everyday resistance (Scott, 1985). Given that it was not realistic to monitor the area completely, it was reasonable that the two sides should reach an informal, implicit compromise. According to a BH villager, FMB officers tolerated villagers cultivating up to the 50 metre line above the dam water level.[5] When they start to cultivate above this line, FMB staff will remove them from the plot. However, this was never made explicit. This was instead a sort of 'equilibrium' between villagers' reclamation attempts and the officers' monitoring activities. This compromise may have been more biased towards those with personal connections with the officers. Nevertheless, we can say that to a certain extent, the villagers' everyday resistance attained this compromise.

PFES and villagers' negotiations

As stated earlier, the villagers felt the FMBs had strengthened their monitoring activities since the introduction of the PFES programme. On the other hand, villagers have received some benefits from the programme. Thanks to assistance from an international NGO, the villagers were allocated one community forest (87 ha) from the FMB in 2014 and were given the right to receive payments from the PFES in return for conserving this forest. In order to play their conservation roles, villagers formed three patrol groups. According to the official instructions, the villagers were to participate in patrol activities on rotation.

However, it seems that the villagers are unwilling to perform this activity. The benefit they have received was far less than they had expected. In 2014, around one million VND (approximately $44 US) was paid to the village. The amount was far too low for the villagers to divide based on households.[6] There are two reasons for the low payment. First, only 26 ha of the total 87 ha was eligible for payment. This was because the forest is a degraded forest with few large trees and some portions completely free of trees.[7] The payment is not based on the land held but on the area of tree cover as evaluated by the FPDF's GIS analysis (discussed shortly). The degraded state of the allocated land reduced the payment amount.

The second reason had to do with the programme's low payment per area (53,000 VND/ha/year in 2014). This may reflect two factors: the watershed area and electricity generation.[8] As explained previously, the total payment from a hydropower plant depends on electricity generation, and the payment rate to FES providers is calculated by dividing the total amount of fees collected through electricity generation by the total forested watershed area. Therefore, if the watershed area is large and electricity generation is small, the amount of payment per hectare to FES providers tends to be small. The payment amount does not relate to how the villagers had contributed labour and played a role in FES provision. It is likely that villagers did not understand how payments worked and thus saw conservation activities as another task required by the state, leading to their minimal contributions.

Given the minimal payment from the allocated community forest, it seems that the PFES made a limited contribution to the villagers' livelihoods. There is also another incident regarding PFES payment in the village. Some sympathetic commune officers and villagers tried to negotiate with the FPDF and the Provincial People's Committee (PPC) over the right to receive PFES payments for the villagers' individual plots in the watershed. It is noteworthy that some commune officers were very sympathetic to the situation BH villagers faced. One officer pointed out that the BH villagers were the ones most affected by the dam's construction but they were paid much less than recipients in other communes. These local officers played an important role in negotiating villagers' rights to PFES payments by providing legitimacy and administrative procedures for the negotiations.

According to the officers, some villagers should be eligible for PFES payments because they planted trees (e.g. acacia and bamboo) on their plots. Though many of these plots were in the pre-relocation village area under FMB ownership, the officers thought the trees, not the land, still belonged to the villagers. For this reason, the villagers should be eligible for payments. They prepared a document and listed 37 households that held such tree-planting plots. Ultimately, the petition was brought to the FPDF and PPC.

As McElwee (2012) argues, the PFES programme rests firmly on the existing state system including the legal framework and norms of the Vietnamese authoritarian state. Naturally, one FPDF officer negatively perceived the petition, stating that the planted trees in the protection forest area belonged to the state, and villagers' customary rights had to follow state law. He added that the villagers misconstrued the law by basing their rights on the fact that they cleared the land and planted trees. Further, FPDF officers in the province changed the payment principle in 2015, such that planted forest (e.g. acacia and rubber) in a production forest area no longer qualified for payment. This decision was justified by the assumption that these tree crops provide less FES than natural forest.

As most villagers' plots are acacia plantations, this policy shift reduced the number of eligible villagers.

Negotiating the meaning of FES

It is generally very difficult for people in Vietnam to negotiate with the state. This does not mean, however, that all petitions and negotiations over forest land rights from villagers are rejected by the state. For instance, a TL commune case in the province demonstrates how the Co Tu ethnic minorities negotiated and won forest land rights, which a state forest enterprise (SFE) had previously removed. The BH village and TL commune cases have similar backgrounds and are both centred on conflicts over forest land allocation. One of the notable differences between them, however, is that the forest area in the BH village is in a more important state location (i.e. hydropower dam site), which is why the PFES was utilized in this case.

In addition, it seems that local negotiation is more difficult in the PFES programme than in other forest land conflicts. In the PFES programme, the information base is automatically utilized to evaluate the forest status and decide whether the payment amount can be established or paid. Hence, the PFES reinforces the influence of official information (including 'scientific' information, as from GIS technology) over field data collected from people. This makes local negotiating power weaker in the PFES than in conventional forest management schemes, which are more of the 'command and control' style, but where the relevant information is fragmented across various government agencies.

Another difference is that the TL negotiations are for villagers' forest land rights, while the BH villagers were negotiating over both the rights and the interpretation of environmental services. The BH villagers' negotiations are worth serious consideration when we think about the very meaning of FES. Whatever the commune officers and villagers intended, this raises questions about who and what provide FES. If the land provides FES, the FMB as the landowner is likely to be the provider and recipient of FES payments. However, this poses some questions: Do the trees contribute to FES and, if so, who are the owners of the trees? Moreover, should 'owners' include the villagers who own the crops? If so, should they receive payment as FES providers?

How to understand the FES was discussed among policy makers and implementers during policy formation and implementation. When the authors asked a FPDF officer about the link between FES and the forest, he admitted that there are no actual studies about this in Vietnam. However, it seemed he assumed that if the forest is 'better' in terms of both quantity and quality, more FES would be expected. It was in this context that he revised the provincial principle of PFES payment to not make payments to acacia plantations in production forest areas because acacia is not a native species and thus creates a degraded forest.

Meanwhile, a national-level policy maker noted the difficulty in defining FES. He told one of the authors that he had a difficult discussion with Ministry of Finance officers regarding how FES payments should operate. During the discussion, the officers continued to ask him what FES are. They understood that FES are use values, but they did not understand how FES are provided by labour in managing the forest; this was because they understood labour contribution in the forestry sector only as labour for tree harvesting.

These instances exemplify the differences in FES meanings among policy makers, implementers, and other potential stakeholders. The FPDF officer and Ministry of Finance officers believed FES derived from the quantity and quality of the forest; thus, FES is a kind of natural environmental rent. On the other hand, the national policy maker's idea includes the providers' labour in the value of FES. Villagers assessed their contribution mostly in terms of their labour contributions since they recognize their benefit as a reward for labour.[9] What hydropower plants expect is a transfer based on the economic value of electricity sold. Thus, the three concepts of FES in the programme – natural environmental rent, labour, and electricity (final services derived from FES) – are all accommodated in the official definition. This sometimes creates contradictions regarding the meaning and value of FES. In all cases, the state monopolizes the authority to define FES under the current PFES programme.

When different rationalities are in conflict, a reasonable way to solve the conflict is through dialogue (cf. Habermas, 1984). In that sense, it seems reasonable that the programme should allow more direct negotiations among the related parties, rather than the current process where the state regulates the whole capitalization process through its mediation mechanism. As Suhardiman, Wichelns, Lestrelin, and Hoanh (2013) suggested, the current programme may ultimately serve as a tool for 'strengthening state control over natural resources' (p. 94). In this regard, the most important message the BH villagers and local officers delivered is that they demanded to have a say in the FES capitalization process. This may also be applied to another important stakeholder in the PFES program: the hydropower company.

Our interviews revealed that even hydropower companies are sometimes willing to talk directly with villagers and commune officers. One company staff revealed that the company is now tackling the problem of unstable water supply from the watershed. In addition to climate change, he suspects that the relocated villagers encroach on the dam watershed area, thereby decreasing the watershed's water holding capacity. However, it is currently unlikely that the state will allow such direct dialogue between the company and the villagers since it monopolizes such political power.

In the policy process of creating natural capital (as in this case), the hybrid characteristics between society and nature (cf. Latour, 1993) or the state of 'human nature' (Castree & Braun, 2001) become more conspicuous.

Under this condition, both discursive and material power structures in the construction of these 'hybrids' are critical for understanding potential social and environmental consequences. The case studies of the PFES programme provide an important lesson when we speak of the so-called market mechanism for natural resource management. In other words, we must consider what mechanisms create this market and how nature is conceptually transformed into capital.

Conclusion

In the era of market triumphalism, the so-called market-based approach was developed to solve environmental problems. Capitalization of nature was one such approach; it was meant to overcome the dilemma between development and the environment, which was why it became popular in some developing countries. This chapter examined the processes and local responses of the capitalization of nature in a programme called Payment for Forest Environmental Services (PFES) in Thua Thien Hue province of Central Vietnam.

Our case study revealed that the state dominated all stages of the programme's capitalization process, particularly in how FES is defined and commoditized. With the FPDF's direction, related information is centralized and compiled at the provincial level by using information technology, such as GIS, which creates an information database to identify, evaluate, and decide to whom, for which forest, and how much payment should be made under the programme. In this sense, it is the state that created this 'market' and 'scientified' FES to strengthen its ability to govern the PFES chain.

Our second finding is that the state's domination in the PFES programme has caused confusion in actual local implementation. FES was interpreted differently among different stakeholders in the programme, which sometimes became a source of misunderstanding and conflict among them. As in the case of the BH village, villagers (and commune officers) often felt the system was unfair in terms of who could receive land rights and the subsequent PFES payment. Some (including local officers) even tried to negotiate FES assessments with state officers based on their partial role in creating the forests. The negotiations tried to point out the contradiction in the official FES definition and to protest the state monopoly in controlling the meaning of FES.

In the PFES programme, however, the chances of winning the negotiation would be smaller because it seems that field-based information (and dialogue) is suppressed by centralized and 'scientified' information processing. When villagers find that petitions and negotiations have failed, they have no other recourse but to conduct everyday resistance through illegal encroachment or shirking conservation tasks. In addition, state domination also precluded direct negotiation or dialogue between villagers and

FES users like hydropower companies. This is in spite of how direct nego-tiation can help both parties obtain better outcomes. By preventing direct discussions, it seems that the state also tries to retain power to mediate between companies and villagers.

These overall results highlight the problem of state monopoly in the creation of natural capital. The introduction of PFES in the study area did not create a sense of fairness in NRD and sometimes caused resistance (either passive or active) by local villagers. Within this process, however, some villagers' questions were very important. They questioned what are included in FES, who and under what mechanism such a 'market' is created, and what kind of idea transforms nature into capital in the process of capitalization. The villagers' struggles exemplified their unful-filled expectation that the principle of payment for environmental services would create a process for their active participation in the coproduction of natural capital.

Notes

1 Also known as payment for ecosystem services.
2 Forest management boards (FMBs) are one of the state-run agencies that are responsible for protecting protection forests and special-use forests.
3 The K coefficient reflects the quality of the forest and other factors affecting forest environmental services. It ranges from zero to one.
4 In addition, contracted owner organizations (such as FMBs) can make subcon-tracts with neighbouring villagers to delegate forest management (Figure 5.1).
5 Villagers tend to start cultivating the lakeside area due to accessibility (they have to cross the dam lake). Their reclamation thus proceeds from the lakeside towards the mountains.
6 Interestingly, the villagers tend to think that the payment from the community forest under PFES should be divided according to their labour contributions. This means that they basically regard the payment as a reward for their labour contributions.
7 The villagers suspected that the FMB might intentionally allocate degraded forest to the villagers, as it is valueless.
8 Another factor is the so-called K coefficient (see footnote 3). In the 2014 payment, this value was fixed to one, which did not affect the amount of payment at all.
9 The VFDP report (n.d.) also notes that

> they often described their actions not in terms of 'protection' but in terms of 'effort.' Just as jobs of similar 'effort' are paid similar wages in the open labour market, households in PFES projects wanted equal payments for similar labour.
>
> (p. 38)

References

Castree, N. & Braun, B. (Eds) (2001). *Social Nature: Theory, Practice, and Politics.* Malden: Blackwell Publishing.

Daily, G. C. & Ellison, K. (2002). *The New Economy of Nature: The Quest to Make Conservation Profitable*. Washington, DC: Island Press.

de Soto, H. (2000). *The Mystery of Capital: Why Capitalism Triumphs in the West and Fails Everywhere Else*. New York: Basic Books.

Epstein, G. A. (Ed.) (2008). *Financialization and the World Economy*. Cheltenham: Edward Elgar.

Ferraro, P. & Simpson, R. (2002). The cost-effectiveness of conservation payments. *Land Economics*, 78, 3, 339–353.

Habermas, J. (1984). *The Theory of Communicative Action, Volume 1: Reason and the Rationalisation of Society*. Boston, MA: Beacon Press.

Latour, B. (1993). *We Have Never Been Modern*. Cambridge, MA: Harvard University Press.

Levitt, K. P. (2013). *From the Great Transformation to the Great Financialization: On Karl Polanyi and Other Essays*. London: Zed Books.

McElwee, P. D. (2012). Payments for environmental services as neoliberal market-based forest conservation in Vietnam: Panacea or problem? *Geoforum*, 43, 3, 412–426.

Pagiola, S., Bishop, J., & Landell-Mills, N. (Eds) (2002). *Selling Forest Environmental Services: Market-based Mechanisms for Conservation and Development*. London: Earthscan.

Pham, T. T., Bennett, K., Vu, T. P., Brunner, J., Le, N. D., & Nguyen, D. T. (2013). *Payments for Forest Environmental Services in Vietnam: From Policy to Practice*. CIFOR Occasional Paper No. 93. Bogor, Indonesia: Center for International Forestry Research.

Polanyi, K. (1957). *The Great Transformation: The Political and Economic Origins of Our Time*. Boston, MA: Beacon Press.

Scott, J. C. (1985). *Weapons of the Weak: Everyday Forms of Peasant Resistance*. Oxford: Oxford University Press.

Suhardiman, D., Wichelns, D., Lestrelin, G., & Hoanh, C. T. (2013). Payments for ecosystem services in Vietnam: Market-based incentives or state control of resources? *Ecosystem Services*, 5, 94–101.

Tercek, M. R. & Adams, J. S. (2013). *Nature's Fortune: How Business and Society Thrive by Investing in Nature*. New York: Basic Books.

VFDP – Vietnam Forests and Deltas Program. (n.d.). *Report on Three Years of Implementation of Policy on Payment for Forest Environmental Services in Vietnam (2011–2014)*. Hanoi: Vietnam Forests and Deltas Program.

Wunder, S. (2005). *Payments for Environmental Services: some nuts and bolts*. CIFOR Occasional Paper No. 42. Bogor, Indonesia: Center for International Forestry Research.

6 Beyond the swans

Cellulose extraction, social mobilization, and environmental transformations in southern Chile

Ricardo Fuentealba and Mariela Ramírez

Introduction

Latin American extractive industries are frequently associated with socio-environmental conflicts where opposite individual and collective interests become evident in relation to uses and socially constructed meanings of the environment (Galfioni, Degioanni, Maldonado, & Campanella, 2013; Seguel, 2010; Sosa, 2005; Walter, 2009). From such opposition emerge social mobilizations that may promote positive transformations with different degrees of success. It is especially relevant to understand under what conditions mobilization contributes to more democratic and inclusive use of the environment.

Chile is not exceptional in regard to the number of environmental conflicts and related social movements. Since the return of the democracy in 1990 the number of such conflicts has risen, reaching 118 cases by 2015 (INDH, 2015). Among these, a particularly important case was an environmental disaster causing death to hundreds of black-necked swans in the Cruces River (Escaida, Jaramillo, Amtmann, & Lagos, 2014; Sepúlveda & Bettati, 2005).

In 2004, a cellulose plant began to operate 56 kilometres north of the city of Valdivia, capital of the Los Ríos region, in southern Chile. The plant belongs to the CELCO-Arauco company, the world's second largest cellulose producer and part of one of the most important economic groups in Chile. As a consequence of a strong opposition to the same project in the coastal settlement of Mehuín, west of Valdivia (Guerra & Skewes, 2004), it was arranged that the plant would discharge its industrial effluent into the Cruces River, even though the river was habitat for one of the most important colonies of black-necked swans in the world and a wetland protected by the Ramsar Convention. Only two months after the plant began operations, the swans began to die (Sepúlveda, 2007). Social and environmental organizations and Valdivia's citizens pointed towards the CELCO-Arauco cellulose plant as the main factor responsible for this environmental disaster.

As a consequence of the swans' death a citizen-based movement emerged coordinated by *Acción por los Cisnes* (APC, Action for the Swans)

(Escaida et al., 2014; Sepúlveda, 2007; Sepúlveda & Bettati, 2005). This was a movement that identified itself as self-funded, self-organized, and made up of individuals not organizations. APC had strong local roots, leading actions against CELCO and its plant and against the state for its passive role during the disaster. The case of the swans' death also gained national and international attention when Greenpeace and a Ramsar commission visited the site of the disaster.

In spite of the strong mobilization, the movement in defence of the swans did not reach its goal of stopping the plant and restoring the wetland to its original state. However, ten years later, the Chilean Supreme Court issued a ruling which ordered CELCO to pay a fine for $5,200 million Chilean pesos (roughly US$7.5 million) recognizing its responsibility for the damage to the wetland. This fine aimed to compensate for the damages caused to the wetland and to create preventive mechanisms of environmental monitoring. The fine indicated that the movement in defence of the swans was correct in accusing this private enterprise of causing the death of the swans.

The movement made visible the lack of transparency and the lack of community involvement in environmental assessment of private projects in Chile. This promoted and contributed to reform of these regulations (APC Webpage; Cordero, 2009; Sepúlveda & Sundberg, 2015; Sepúlveda & Villarroel, 2012). In addition, the attitudes of people in Valdivia changed (Morales Aguirre, 2012; Oñate, 2009), and the practices of private enterprises were affected (De la Maza, 2012; Sepúlveda & Sundberg, 2015).

This chapter explores transformations generated by this movement. In general, one of the less researched topics concerning socio-environmental movements relates to understanding their outcomes (Giugni & Grasso, 2015; Rucht, 1999). This kind of study is crucial to learn under which circumstances a movement's agenda is met, but also to understand what are the unintended transformations associated with a mobilization. We argue that the movement defending the swans generated impacts beyond their demands to clarify the causes of the swans' death. These wider impacts are explained by both, the movement's organization and its strategies, and the broader context within which the mobilization was developed.

After this introduction, the chapter is divided into four sections. In the next section we present our research approach to the case in Valdivia. Then we present the results of the study in two sections: first we address the characteristics of the movement and its context, and then we describe the impacts and reflect on the factors leading to these impacts. We then conclude.

Understanding the impacts in Valdivia

The case of the conflict for the defence of the swans has been widely studied (Delgado, Marín, Bachmann, & Torres-Gomez, 2009; Ehrnström-Fuentes,

2015; Leal & Negrón, 2012; Oñate, 2009; Sepúlveda, 2007, 2016; Sepúlveda & Mariángel, 1998; Sepúlveda & Rojas, 2010; Sepúlveda & Villarroel, 2012; Villarroel, 2014) and there are even previous studies on some of its consequences (De la Maza, 2012; Morales Aguirre, 2012; Sepúlveda & Sundberg, 2015; Sepúlveda & Villarroel, 2012). Our study is part of this latter body of research, centred specifically on the political impacts of this movement. In this section, we describe the approach used along with some methodological concerns.

Following the recommendations of Tilly (1999) and Amenta and Young (1999), we explored explicitly the impacts that go beyond the movement's demands. This entails analysing both the movement's objectives and also the myriad shifts found as a consequence of, for instance, their strategies. Also, based on the discussion about how to name this set of transformations associated with a social movement – either impacts, consequences, results or effectiveness – we will speak about impacts, following Amenta and Young (1999) who argue that 'it may be possible for a challenger to fail to achieve its stated programme – and thus be deemed a failure – but still to win substantial collective benefits for its constituents' (1999: 25). This is fundamental for the case we are analysing given the movement's features and demands but also the characteristics of Chilean institutions.

Another issue refers to how to analyse and interpret the results. In other words, what dimensions best explain the impacts we found and how we make sense of them. A possible framework could be the work of Gamson (1975) when he distinguishes agenda, incentives, strategies and organization to account for the effectiveness of movements. However, we agree with Giugni (1999) in regard to the necessity of addressing both internal and external characteristics, therefore we analysed factors from the movement itself and from its wider context. We analysed three set of variables. First, we examined the characteristics of the socio-environmental movement in Valdivia. In particular, we followed Amenta, Caren, Chiarello, and Su (2010) and described the movement's strategies, organization, and context. Second, we analysed the political impacts that relate to this movement. This entailed an exploration of specific political phenomena (policies, political institutions, political processes, power relations) that linked to the movement's action and context. And third, we analysed the concrete mechanisms through which the movement influenced these impacts. The latter, we believe, is fundamental in our case to overcome the allocation of causality problem identified by Giugni (1998, 1999), as these practices and processes are what allows us to link the movement and its context, with the impacts.

To do this, we followed a qualitative strategy. We conducted ethnographic work in Valdivia, located 849 kilometres south of Santiago, with a population of 140,000 and the capital of the Los Ríos Region. We participated in several community events and activities related to environmental initiatives. We also interviewed key informants with different opinions and levels of participation in the mobilization, among others: men and women who were spokespersons during the conflict against CELCO and who led the organization of the movement; academics from the Universidad Austral de Chile, the main educational institution in Valdivia; social and community organization representatives who participated in several activities to defend the swans; and local authorities and workers in the public sector. The interviews were audio-recorded, transcribed and codified using Atlas.ti. In addition to this primary information, we collected secondary material. Among these, previous research on the case of the conflict, local and national press (both printed and digital), and technical documents of public use (territorial plans, local policies, etc.). These were also processed using Atlas.ti, codified along with the interviews.

Finally, we want to assert two advantages for this research based on the span of time since the conflict. First, we believe that, after more than a decade, it is feasible to delve into the impacts related to this social movement. Analogously, nobody can deny the effects of the French Revolution or the French student revolt in May 1968; however, such effects were hardly accounted in 1790 or June 1968, respectively. And second, a decade after the most critical moments of the conflict, local actors are more reflective about their participation in the movement and its significance. In this sense, how the interviewees interpreted their past and the effects of the mobilization enriched enormously our own explanation of the phenomena.

The movement in defence of swans

In this section we describe some key process features of the movement in Valdivia, detailing in particular its organization, composition, strategies, and the context in which it acted. We understand the movement of Valdivia as the set of individual and collective actors involved in actions to stop the environmental damage of the Río Cruces Wetland that led to the death of the swans. These organized around the citizen group Acción por los Cisnes (APC, Action for the Swans). APC was created on 2 November 2004, the day the first public citizen assembly was held, with the aim to draw attention to the environmental disaster. The first denunciations were prepared (APC Webpage) and APC was established as an interlocutor between the citizen's movement and public institutions. According to a statement in which APC described its objectives, agenda and functioning, its independence and transversality was asserted, which ultimately delivered APC its legitimacy to act (Oñate, 2009).

APC had a core of people with highly specialized knowledge from experience with other social movements and community work; from the world of arts and culture; and from different academic backgrounds in disciplines such as biological sciences, social sciences, medicine, legal studies, and so on. As one of their spokespersons described:

We could resolve internally almost all of the things we had to do to move a citizen's movement: we had convening power, good social leaders [...] We basically resolved all technical, legal and scientific support ...

Seven of its representatives were spokespersons, with roles that depended on specific capacities required for the particular situations they faced, sharing and debating with citizens in open assemblies organized weekly. These assemblies were held in public spaces and gave the mass character to the movement's messages, linking their demands to the community.

Due to these actions, the movement transitioned from a mostly environmental organization towards wider and more cross-sectional support in social and economic terms, transforming itself into a movement of Valdivia. Such diversity can be appreciated with the involvement of the private sector, notably the tourism and commerce sector. APC also gathered support from neighbourhood councils, rural and indigenous communities, and cultural and student organizations. Public services supported the movement, either directly through public announcements or indirectly by providing technical documents about the scope of the environmental disaster.

The demands of the movement were heterogeneous. APC for instance declared as their objective to 'stop the causes of the severe ecological deterioration affecting the Cruces River along with the risks this situation represents for people's health'. As a spokeswoman summarized, the central theme of APC was 'to stop the disaster and restore the wetland'. Besides these, APC also declared a concern to demonstrate the consequences of the disaster to Valdivia's identity.

Nonetheless, the movement had as the main object the defence of the dying black-necked swans, animals with important historical meaning linked to the identity of Valdivia and the region. This element was crucial to understanding their capacity to attract support. Many interviewees argued that the figure of the swans was the main element motivating people to attend the different activities carried out by APC (Sepúlveda, 2007, 2016). Some people were motivated by dying swans falling into their backyards, others worried about the city's water health, and others were affected by a disaster that threatened their identity. Ultimately, what mobilized the people in Valdivia was the desire to understand the causes of the swans' death at a moment of high uncertainty, hence, to stop the contamination of the Cruces River wetland and restore it. We do not make a distinction between restoring the wetland and saving the swans since both are interrelated.

A critical feature was the movement's strategy. APC declared that its work was based on 'demanding transparent information regarding what happened [in the Cruces River]' and therefore applying 'rigorously (...) the environmental legislation'. In this sense, one of the most important elements of the movement according to the interviewees was their explicit intention to use all institutional channels available at the time. As a spokeswoman stated: 'the movement for the swans always works with the institutional rules, [you] always (...) can use the rules in your favour'. An example of the latter is what happened after the *Consejo de Defensa del Estado* (CDE, Defense Council of the State) demanded CELCO to pay for the restoration of the wetland, along with compensating for the damage to Chile's environmental heritage. APC actors worked alongside CDE to make CELCO pay a fine and design a plan to restore the wetland and develop community projects.

Another important point of the movement's strategy refers to its mobilization activities. APC used non-violent means to protest which, according to participants, involved stipulating strategies for public demonstrations and sanctioning those who caused disturbances. Along with that, the marches were convened as family meetings, scheduled when small children and young students could attend. Besides marches, APC carried out many other public activities to mobilize the city of Valdivia, such as communication campaigns, gatherings with authorities, fieldwork at the Cruces River wetland, and environmental education activities, among others. These different initiatives were a huge advantage for the movement:

> they were clearly a contribution to the discussion (...) (along with) that familiar tone, more playful. In the march we forbid painting walls or throwing papers, something that was fulfilled sacredly (...) We had people controlling that; even the political banners (had to go) to the end of the march ...

Many activities were influenced by people belonging to the culture and the arts world, who transformed public demonstrations into cultural performances. This is also recognized as a key feature of this case, as they designed strategies to deliver the movement's messages more creatively.

The features described earlier correspond to the interior of the movement – but what were some of its external dimensions? A critical contextual feature is the profound centralization of Chilean regional administration, including its development and environmental institutions. This is important for the case not only for the lack of participation in processes of environmental assessment (Sepúlveda & Bettati, 2005; Sepúlveda & Mariángel, 1998; Sepúlveda & Rojas, 2010), but also because the regional authorities lacked administrative capacity to respond accordingly. The death of the swans erupted at a critical period in which the Los Ríos Region gained autonomy, a milestone in the history of regional governance in Chile (Escaida, Miranda, & Vega-Duarte, 2016).

Another important aspect of the movement's context is the interest of Chile to enter the Organization for Economic Cooperation and

Development (OECD). As part of the country's application, the OECD analysed its environmental institutions and concluded that Chile has enormous deficiencies in relation to its normative framework, its environmental audits, planning capacities, and assessment processes, among others (OECD, 2005). This study ultimately put in question the environmental framework of the country, already criticized by several socio-environmental movements – another feature of the context. According to Sepúlveda and Villarroel (2012, p. 191) 'the Valdivian movement is better understood in terms of its material and symbolic connection to the larger process of consolidation of the 'internal environmental demand' that evolved as a result of the crisis of legitimacy of Chilean environmental institutions', relating this case to dozens of other environmental movements with similar experiences. Therefore, both the negotiation of Chile entering the OECD, which it did in 2010 becoming the first South American country in the organization, and the number of previous conflicts gave more visibility to the movement in Valdivia.

Lastly, the media played a role as a political tool. There are several analyses of the behaviour of Chilean media during this socio-environmental conflict (Ehrnström-Fuentes, 2015; Leal & Negrón, 2012; Oyarzo, 2014), mainly addressing their relation to big enterprises. In the case of Valdivia, the two main newspapers in Chile took different positions in relation to CELCO, giving this private actor high visibility throughout the conflict to present their defensive arguments. Even more, the media worked instrumentally to legitimize certain views of the state and private enterprises and not those of APC and local citizens (Ehrnström-Fuentes, 2015; Oyarzo, 2014). Thus, while the movement's demands were underrepresented and appeared as an actor of the debate when these media recognized them as valid, the media advanced other arguments such as the positive impacts of the cellulose plant for the economy of the region as general facts. Having described the movement in defence of the swans, now we turn to analyse the movement's impacts and argue that these have reached beyond the scope of the movement's immediate demands.

Going beyond the swans: what impacts and why

To understand why we argue that the movement had impacts 'beyond the swans', we first attend to changes based on the movement's objectives. This is important because, as several interviewees and spokespersons stated, the movement 'was not successful' and even 'failed completely' on its demands, mainly because the source of pollution (the cellulose plant) was not eliminated. Furthermore, the fine that CELCO paid was criticized by some participants of the movement, considering that the set of projects and investments financed are insufficient in relation to the scale of the disaster and the damage to the region's environmental heritage.

As a spokeswoman declared:

> The court ruling did not order material reparation measures, but other kinds such as studies, diagnostics, or [the creation of] a research center. These elements are relevant to creating different practices around the wetland but do not repair the disaster. [The wetland] is far from being in the state it was before the conflict.
>
> (Radio Universidad de Chile, 2015)

Independent of this, APC declared among its achievements having provided evidence of 'CELCO's non-compliance and the deficiencies of the government's audit', along with 'uncovering the serious limitations of [Chile's] environmental institutions'. Even more, APC states that 'the 'swans' case' of Valdivia has marked a turning point in Chile's environmental institutions, initiating a discussion about our environmental rights and our work as citizens'. Although not part of the movement's stated objectives described in the last section, many interviewees stated that the participants in the movement started to be more conscious about the need for more profound actions to alleviate this disaster. As a spokesman recounted:

> some people arrived to this social movement solely to save the swans, but realized that in order to do that they must also save the river, and to save the river they should close the pulp mill, and to close it they should change the environmental institutions, and to change these, they have to change the whole political constitution of the [Chilean] State.

Among such profound changes, it has been argued that the movement influenced the reform of Law N° 19.300 on the General Bases of Environment of 2009 (Sepúlveda & Villarroel, 2012). Others argue instead that this reform is mainly a consequence of the study on the Environmental Institutions of Chile conducted by the OECD (2005) (Tecklin, Bauer, & Prieto, 2011). From our point of view, the 2009 Environmental Reform is a product of both this movement and the OECD study, fuelled also by the increasing public demand and some further critiques of the previous Environmental Law by conservative media (Oyarzo, 2014).

There are two mechanisms that link the movement with this reform – besides being considered 'the last straw' of a process of accumulating environmental conflicts, as Sepúlveda & Villarroel (2012, p. 183) argue. First, throughout the conflict and based on the positive reaction of public opinion, the movement gained support from different political actors including from the legislative sector.[1] These actors not only started to question CELCO's actions during the conflict but also openly criticized the capacity of the previous environmental regulations, hence pushing forward shifts on these regulations. And second, the movement's strategy to use all institutional channels. Using the existing environmental framework effectively, allowed the movement not only to have the opportunity to influence policy changes but also to show the weaknesses of these institutions.

Beyond this change in the environmental institutions at the national scale, we think there are spaces to push other institutional changes especially at the regional level, although constrained by the aforementioned centralization of Chile's political institutions. For instance, the regional model of representation remains based on appointments from the central level in Santiago, which ultimately render unlikely shifts in formal regional administrative rules. However, we think there are some changes that manifest more informally, mainly regarding how different actors relate to formal institutions and how power has been re-balanced in the region. We centre our attention on these.

After the conflict the legitimacy of regional actors to conduct environmental assessment processes came under question. In this sense, people realized the need to include other actors and to participate more widely in such processes. We witnessed an increase of participation by civil society in the few environmental institutions that allow their involvement, especially in the assessment of projects based on the exploitation of natural resources. An example of this is the recent citizen response to an investment project to generate wind energy in the coastal part of the Valdivian forest. The project was presented to the Environmental Assessment Service for official evaluation. In response there was a strong communication campaign led by civil society organizations and including some previous spokespersons of the movement for the swans. The campaign tried to mobilize people to use all institutional channels to stop the project, which translated in more than five thousand observations during the citizen participation stage (Biobío Chile 2016), which impeded the project's process of environmental assessment.

What we argue here is that after the conflict in defence of the swans, the process of decision making in Valdivia regarding projects using natural resources involved a much wider and diverse set of actors. One of the main reasons for this was the local roots of the movement which developed strong leadership by people in the city and that has continually been involved in cases of environmental conflicts. Another reason is the learning advanced by the movement's strategy. As mentioned, this is particularly important with respect to using all the rules provided by the assessment processes, defending natural resources in the city and its environment through all institutional paths available, and positioning environmental organizations as valid actors.

Some interviewees associated the movement with an increased level of citizen engagement in public matters, particularly on environmental-related issues. Los Ríos Region occupies the second place (among 15 regions) in the number of organizations per thousand habitants,[2] with an overrepresentation of those dedicated to environmental protection and conservation

(Centro UC de Políticas Públicas, 2016). Other interesting results come from a recent discussion for constitutional change in Chile.[3] According to these, Valdivia is the second Chilean city in participation per capita,[4] and is the only city that prioritizes the respect and conservation of nature as the value that should inspire the new constitution.

There are many different meanings of concepts such as 'environment' or 'nature', from a purely utilitarian one to a more ecological meaning (Purdy, 2015). In the case of Valdivia, a study on the identity of the region concluded that its environment is one of the cultural axes of its identity (Gobierno Regional Los Ríos, 2010b). The study, nonetheless, does not explain this feature. There is some incipient research on how the relation between Valdivian citizens and their environment has shifted – and on the role of the movement for the swans on this change (Sepúlveda, 2016). Relatedly, we argue that the same can be said about urban wetlands. According to some interviewees, these spaces were perceived negatively by local communities before the conflict, naming them pejoratively as '*hualves*' and characterizing them as sites without value, wastelands that were ultimately foci of infections. Nonetheless, a direct impact of the movement defending the swans was the radical transformation of how these urban wetlands were perceived by the people in Valdivia, not only in regard to their aesthetic features but also integrating these to their communities and the city. A spokeswoman described this change:

> Because of this disaster we know where we live. We can see our city from another perspective. We can realize that we are inside the wetland: is not that the wetland is north of Valdivia, but Valdivia is inside the wetland. We are inhabitants of the wetland.

Hence, the urban wetlands in particular and environmental issues more generally are strongly incorporated into the public agenda. The discourse about nature was redefined and empowered, leading to a myriad of new practices with more integrated goals. There are many examples of this in local policies and planning processes. One of the most meaningful is the incorporation of the disaster into the Regional Development Strategy. This Strategy is based on many scenarios for the future of the region based on the vision of local communities, one of which was called 'the return of the swans'. This scenario has a strong component of regional development based on sustainable economic activities, mainly tourism. It is fuelled by ideas of preservation of natural resources, biodiversity and cultural activities; linking nature to cultural heritage; increasing protected areas, biological corridors; reducing territorial conflicts as a result of better water governance; and efficiently applying environmental management instruments (Gobierno Regional de Los Ríos, 2010a).

Conclusion

At the beginning of the chapter we asserted that environmental conflicts and mobilization in Latin America are associated with the region's dominant development strategy of natural resource exploitation. There is a significant body of research about the emergence of socio-environmental movements contesting this development path, characterizing their adversaries, composition, strategies, etc. Nonetheless, there is much less information about the kind of transformations these movements have generated and this chapter contributes to that research.

The socio-environmental movement that arose in Valdivia as a response to the disaster in the Cruces River that caused the death of thousands of swans, was a very particular one. It had strong local rootedness, a cross sectional and diverse composition, multidisciplinary resources, and a strategy to use all available institutional means. It had an ability to convene broad engagement in its activities, it was explicitly and deliberatively non-violent, and it presented itself to the public as open and democratic, leading ultimately to an important identification by local people.

As we have described and argued in this chapter, both the movement's internal features and its contexts helped push forward a set of social transformations having results beyond the conflict. Along with other impacts of this movement already researched such as the 2009 environmental reform or the resignification of the figure of the swan as a local icon, we have identified the crucial role of this case in shifting the power balances in the territory, increasing community engagement and participation in environmental projects, and redefining the significance of spaces such as urban wetlands.

We conclude then by pointing out that socio-environmental movements can have a myriad of potential impacts to be explored. Although such changes might not resolve the dilemmas about nature or impede the emergence of environmental conflicts, these can give specific traces of new forms of governance in a city and/or its environment. Therefore, we argue that social movements that contest specific natural resource projects can lead to positive environmental transformations, bringing forward more democratic forms of decision-making and resource governance.

Notes

1 Among the political actors raising criticism regarding environmental institutions were the President of the Parliament's Environmental Commission, Leopoldo Sánchez, and Congressman Guido Guirardi and Enrique Jaramillo.
2 The first being the Araucania Region, which has the largest indigenous population and number of indigenous communities in the country.
3 Since 2016, a process to change the national constitution is being held. The first activity was a participatory process at the local and territorial levels. Results from this stage can be found in Proceso Constituyente de Chile.
4 Coyhaique is another city that suffered a strong environmental conflict between 2011 and 2014 against a hydroelectric project and is the first.

References

Amenta, E., Caren, N., Chiarello, E., & Su, Y. (2010). The political consequences of social movements. *Annual Review of Sociology* 36, 287–307.

Amenta, E. & Young, M. P. (1999). Making an impact: Conceptual and methodological implications of the collective goods criterion. In M. Giugni, D. McAdam, & C. Tilly (Eds) *How Social Movements Matter*, pp. 22–41. Minneapolis, MN: University of Minnesota Press.

APC Webpage. Acción Por los Cisnes. URL: www.accionporloscisnes.org. Retrieved on 10 November 2016.

Biobío Chile. (2016). Recolección de observaciones contra Central Eólica Pililín culminó con más de 5.000 firmas [Press release]. Retrieved 20 October 2016, from www.biobiochile.cl/noticias/2016/04/20/recoleccion-de-observaciones-contra-central-eolica-pililin-culmino-con-mas-de-5-000-firmas.shtml.

Centro UC de Políticas Públicas (2016). *Mapa de las organizaciones de la sociedad civil. Primer informe de resultados del proyecto Sociedad en Acción.* Santiago: Centro de Políticas Públicas, Pontificia Universidad Católica de Chile.

Cordero, Luis (2009). La ruta del rediseño de la institucionalidad ambiental. In E. Aliste & A. Urquiza (Eds) *Medio ambiente y sociedad. Conceptos, metodologías y experiencias desde las ciencias sociales y humanas.* Santiago: Ril Editores and Universidad de Chile.

De La Maza, G. (2012). Responsabilidad social empresarial, política e internacionalización. El caso del 'conflicto de los cisnes' en Valdivia, Chile. *Apuntes, Revista de Ciencias Sociales.* 39, 70, 167–202.

Delgado, L., Marín, V. H., Bachmann, P. L., & Torres-Gomez, M. (2009). Conceptual models for ecosystem management through the participation of local social actors: the Río Cruces Wetland conflict. *Ecology and Society*, 14, 1, Online, URL: www.ecologyandsociety.org/vol. 14/iss1/art50/.

Ehrnström-Fuentes, M. (2015). Production of absence through media representation: A case study on legitimacy and deliberation of a pulp mill dispute in southern Chile. *Geoforum*, 59, 51–62.

Escaida, J., Jaramillo, E., Amtmann, C., & Lagos, N. (2014). *Crisis socioambiental: El Humedal del Río Cruces y el Cisne de Cuello Negro.* Valdivia, Chile: Ediciones UACh, Colección Austral Universitaria de Ciencias Sociales, Artes y Humanidades.

Escaida, J., Miranda, J. C., & Vega-Duarte, F. (2016). La Región de Los Ríos como proyecto político subnacional: una mirada al proceso de una región nueva. *Revista de Direito da Cidade*, 8, 2, 538–567.

Galfioni, M., Degioanni, A., Maldonado, G., & Campanella, O. (2013). Conflictos socioambientales: identificación y representación espacial. Estudio de caso en la ciudad de Río Cuarto (Argentina). *Estudios Geográficos* Vol. LXXIV, 275, 469–493. DOI: 10.3989/estgeogr.201317.

Gamson, W. A. (1975). *The Strategy of Social Protest.* Belmont, CA: Wadsworth.

Giugni, M. (1998). Was it worth the effort? The outcomes and consequences of social movements. *Annual Review of Sociology*, 24, 1, 371–393.

Giugni, M. (1999). Introduction. How social movements matter: Past research, present problems, future developments. In M. Giugni, D. McAdam, and C. Tilly, (Eds) *How Social Movements Matter*, pp. xiiix–xxiii. Minneapolis, MN: University of Minnesota Press.

Giugni, M. & Grasso, M. T. (2015). Environmental movements in advanced industrial democracies: Heterogeneity, transformation, and institutionalization. *Annual Review of Environment and Resources*, 40, 337–361.

Gobierno Regional Los Ríos. (2010a). *Estrategia Regional de Desarrollo*. Valdivia: Equipo Técnico del Gobierno Regional de Los Ríos.

Gobierno Regional Los Ríos. (2010b). *Estudio para el Fortalecimiento de la Identidad de la Región de Los Ríos*. Valdivia. Retrieved 1 November 2016 from www.goredelosrios.cl/cultura2/wp-content/uploads/2016/02/Estudio-para-el-Fortalecimiento-de-la-Identidad-de-la-Regi%C3%B3n-de-Los-R%C3%ADos-Versi%C3%B3n-2015-Gobierno-Regional-de-Los-R%C3%ADos.pdf.

Guerra, D. E. & Skewes, J. C. (2004). ¿Qué Fue lo que Resultó? Mehuín (Chile, Décima Región) y su Defensa del Medio Ambiente: Proyecciones para la Protección Comunitaria de los Recursos Patrimoniales. In *Actas del 5º Congreso Chileno de Antropología. Tomo*. 1, 594–602. V Congreso Chileno de Antropología. Colegio de Antropólogos de Chile A. G, San Felipe.

INDH (2015). *Mapa de Conflictos Socioambientales en Chile. Versión 2015*. Santiago, Chile: Instituto Nacional de Derechos Humanos.

Leal, F. & Negrón, M. (2012). Tensiones socioambientales y rol de los medios regionales de comunicación en la formación de debate público: dos casos emblemáticos para la institucionalidad ambiental chilena (CELCO y Barrancones). *Revista Austral de Ciencias Sociales* 22, 25–42.

Morales Aguirre, B. (2012). La defensa del medioambiente y la construcción de ciudadanía: reflexiones en torno a un 'movimiento ciudadano' en la ciudad de Valdivia, Chile, *Amérique Latine Histoire et Mémoire. Les Cahiers ALHIM*. 24, Online. Retrieved 15 December 2016 from http://journals.openedition.org/alhim/4421.

OECD. (2005). *OECD Environmental Performance Reviews – Chile*. Paris: Organization for Economic Cooperation and Development.

Oñate, B. (2009). *Construcción Social del Medioambiente El Movimiento Ciudadano Acción por los Cisnes Caso CELCO-Valdivia*. Tesis para optar al Título Profesional de Antropóloga Social, Facultad de Ciencias Sociales, Departamento de Antropología, Universidad de Chile, Santiago.

Oyarzo, M. (2014). *El imaginario social construido por la prensa chilena sobre la contaminación del río cruces en valdivia*. Tesis doctoral, Facultad de Ciencias de la Comunicación, Departamento de Medios, Comunicación y Cultura. Universidad Autónoma de Barcelona.

Proceso Constituyente de Chile. www.unaconstitucionparachile.cl. Retrieved on 15 December 2016.

Purdy, J. (2015). *After Nature: A Politics for the Anthropocene*. Cambridge, MA: Harvard University Press.

Radio Universidad de Chile (2015). *Vocera del Movimiento Acción por los Cisnes: 'No podemos decir que esto nunca más va a ocurrir en el país'*. Retrieved on 20 October 2016, from: http://radio.uchile.cl/2015/04/19/vocera-del-movimiento-accion-por-los-cisnes-no-podemos-decir-que-esto-nunca-mas-va-a-ocurrir-en-el-pais/.

Rucht, D. (1999). The impact of environmental movements in western societies? In M. Giugni, D. McAdam, & C. Tilly (Eds) *How Social Movements Matter* (pp. 204–224). Minneapolis, MN: University of Minnesota Press.

Seguel, Andrés (2010). Nuevas formas de agencia social: de la visibilidad de los conflictos a la globalización de los objetos medioambientales. In: E. Aliste and A. Urquiza (Eds) *Medio ambiente y sociedad: conceptos, metodologías y experiencias desde las ciencias sociales y humanas.* Santiago, Chile: RIL editores. ISBN 978-956-284-727-8.

Sepúlveda, C. (2007). *Ciudad de Papel.* Valdivia, Chile: Jirafa Films.

Sepúlveda, C. (2016). *Swans, Ecological Struggles and Ontological Fractures: A Posthumanist Account of the Río Cruces Disaster in Valdivia, Chile.* PhD Thesis, The Faculty of Graduate and Postdoctoral Studies (Geography), The University of British Columbia.

Sepúlveda, C. & Bettati, B. (2005). El Desastre Ecológico del Santuario del Río Cruces: Trizadura Institucional y Retroceso Democrático. *Revista Ambiente y Desarrollo*, 20(3)–21(1), 62–68.

Sepúlveda, C. & Mariángel, P. (1998). La Legitimidad del Sistema de Evaluación de Impacto Ambiental Puesta en Juego: el Caso de la Planta de Celulosa Valdivia. *Ambiente y Desarrollo*, 14, 2, 6–17.

Sepúlveda, C. & Rojas, A. (2010). Conflictos Ambientales y Reforma Ambiental en Chile: Una Oportunidad Desaprovechada de Aprendizaje Institucional Sobre Participación Ciudadana. *Revista Ambiente y Desarrollo*, 24, 2, 15–23.

Sepúlveda, C. & Sundberg, J. (2015). Apertura Ontológica, Multiplicidad y Performación: Explorando una agenda post humanista en Ecología Política a partir del desastre del Río Cruces en Valdivia. In B. Bustos, M. Prieto, & J. Barton (Eds) *Ecología política en Chile: naturaleza, propiedad, conocimiento y poder.* Santiago, Chile: Editorial Universitaria.

Sepúlveda, C. & Villarroel, P. (2012). Swans, conflicts and resonance: The role of local movements in the reform of Chilean environmental institutions. *Latin American Perspectives*, 39, 4, 181–200.

Sosa, E. (2005). Los Conflictos socio ambientales en la provincia de Mendoza: Marco conceptual. In A. Scoones and E. Sosa (Eds) *Conflictos socio ambientales y políticas públicas en la provincia de Mendoza.* Mendoza, Argentina: Oikos Red Ambiental.

Tecklin, D., Bauer, C., & Prieto, M. (2011). Making environmental law for the market: the emergence, character, and implications of Chile's environmental regime. *Environmental Politics*, 20, 6, 879–898.

Tilly, C. (1999). Conclusion. From interactions to outcomes in social movements. In M. Giugni, D. McAdam, & C. Tilly (Eds) *How Social Movements Matter* (pp. 253–270). Minneapolis. MN: University of Minnesota Press.

Villarroel, P. (2014). *Sociedad del riesgo y comunicación social de la ciencia: Apropiación social del conocimiento científico relevante en el marco de conflictos ambientales. El caso de un desastre ecológico en el sur de Chile.* Tesis Doctoral, Facultad de ciencias de la Comunicación, Departamento de Periodismo y Ciencias de la Comunicación. Universidad Autónoma de Barcelona.

Walter, M. (2009). Conflictos ambientales, socioambientales, ecológico distributivos, de contenido ambiental. Reflexionando sobre enfoques y definiciones. *Boletín ECOS*, 6, Febrero-Abril. Madrid, Spain: Centro de Investigación para la Paz (CIP-Ecosocial).

7 'No oil in our soil!'

Shifting narratives from commodities to the commons in Iowa, USA

Angie Carter and Ahna Kruzic

Introduction

> I'd like to sum it up in the sentence or two that my mother-in-law said the other day. She looked out her window and said to me 'I just can't stand to see that land over there torn up for this pipeline. I've worked so many years taking care of it. What is the real benefit of it?' It's not just us we're concerned about. We're thinking of future generations. We have to have faith, and I have to believe that sooner or later, down the line, this pipeline will be stopped.
>
> (Pam Alexander, speaking on her Iowa farmland, 22 July 2016, in Vanderpol 2016)

The Bakken Pipeline Resistance Coalition[1] – a grassroots group that began organizing in Iowa, United States against the Dakota Access pipeline – formed during mounting tensions in the state concerning agricultural water pollution and rights to water as a public good. Early opponents of the pipeline argued the project presented risk to Iowa's farmland, a source of identity for many across the state. In this chapter, we focus our study on the how the Coalition created a narrative transforming arena through specific narrative-shifting processes. The Coalition first began organizing against the pipeline in late summer 2014, construction of the pipeline began in summer of 2016, and the pipeline began carrying oil in early summer 2017. As illustrated in the opening quote from Iowa farmland owner, Pam Alexander, the pipeline resistance created an opportunity for important shifts in cultural narratives about Iowa farmland. In this chapter, we identify three strategic processes used by the Coalition as they reframed resistance narratives from questions of local farmland owners' rights to the future of the commons.

Land has long been used as a mechanism to maintain power. Neoliberalism dominates the political and economic realities of land; the privatization of land and deregulation of its management are defining features of the centralization of power in the United States. To be successful, neoliberalism requires the dismantling and privatization of the commons (Federici & Caffentziz, 2013). This neoliberal ideology shapes cultural narratives about agriculture by prioritizing profit and yield in both the value of farmland and the value of the farmer. The dominant methods of agricultural production practiced by the majority of Iowa's farmers are characterized as productivist agriculture, described as promoting the intensification of technology, concentration of farmland, and specialization of production rather than diversification (Trauger, 2001, p. 154).

The Coalition, as a narrative-shifting arena, provided space for collaborative interactions among unlikely allies who re-evaluated, re-defined, and re-constructed narratives about Iowans' relationships to land, water, and each other. Narratives are our focus in this chapter because narratives are key to changes in individual and cultural identity (Loseke, 2007). We argue that the Coalition's framing of resistance around the commons challenged the dominant productivist narrative and was successful in creating a new space for claims-makers resisting extractive energy projects. In elevating the commons, the Coalition's narrative work extended beyond a defensive reaction to crisis (Zibechi, 2012). Instead, this work began to plant 'the seeds, the embryonic form of an alternative mode of production' in the making (Federici & Caffentziz, 2013, p. 95). Through analysis of the Coalition's narrative reconstructions, we identify three strategic processes used by the Coalition to create important shifts in cultural narratives about agricultural land in Iowa: creating space for new claims-makers, elevating the commons, and aligning values with regional and national campaigns against extractive energy.

Our findings suggest that while unsuccessful in stopping the construction of the pipeline, the Coalition has been successful in creating a collective, grassroots space for continued organizing which has reconstructed polarizing narratives about property rights into a more inclusive, broader-reaching narrative centring on collective legacy and public goods. Further, the Coalition united Iowans from across existing rural/urban and political divides and aligned with larger regional and national anti-extractive movements. In response to McAdam and Boudet's (2012) critique that too often social movement analysis focuses only on stories of successful mobilization, we share an ongoing story of smaller wins defined in terms of narrative shifts. These shifts, if sustained, may continue to create opportunities to centre justice and the commons in resistance to extraction. We hope our analysis of the Coalition's narrative shifting strategies will inform future grassroots resistance efforts, especially those in agricultural landscapes where narratives about maximizing profit per acre dominate individual and cultural identity construction as well as state policy.

We will first situate the Coalition's work within the Iowa landscape – one defined by neoliberal narratives of agricultural extraction prioritizing yield and profit. We will then discuss the methodology of our narrative analysis, including our positionality as researchers and participants in the resistance. Next, we will define the three narrative shifting processes identified through our narrative analysis and conclude with what we have defined as wins – shifts in messaging and collective action – that we hope will be helpful to others engaged in the hard work of grassroots action.

Iowa's agricultural landscape and narratives

The history of land use in Iowa is important to understanding how dominant cultural narratives shaped the emergence of the Bakken Pipeline Resistance Coalition. Narratives are 'our most elemental human genre of communication and meaning-making, an essential way of framing the order and purpose of reality' (Smith, 2003, pp. 151–152). Cultural narratives circulate stories sharing expectations, rights, responsibilities, and norms and are important to the study of social change because they are sites for intervention, redefinition, and shifts in power (Loseke, 2007).

Power, in the context of landscapes, is the ability to define and control knowledge and meaning while mobilizing support (Greider & Garkovich, 1994, p. 17). In practice, racialized and patriarchal social norms in dominant cultural narratives about farmland present challenges for collaboration among the many demographic groups along the pipeline route – indigenous people, students, farmers, faith-based groups, urban and rural residents. We have focused our study on these cultural narratives because they offer important frameworks for understanding how individuals construct meaning in relation to power, land, and resources, and how they redefine symbolic codes to construct new meaning in relation to the commons.

Loseke's (2007) theory of narrative construction, rooted in Berger and Luckmann's (1967) theory of social construction, explains how meaning-making works at multiple levels to re-enforce dominant and hegemonic cultural stories. This power translates to the ability to impose a definition of the physical environment – a definition serving the interests of particular institutions or people (Grieder & Garkovich, 1994, p. 17). The power relationships involved in defining and controlling landscapes and identity are critical to understanding 'how physical 'nature' is managed and altered according to the dominant set of social values within the culture' (Joubert & Davidson, 2010, p. 9). Expectations of corn and soybean farmers in rural Iowa to value yield and maximize profit become embedded and institutionalized over time as cultural stories (Hopper, 2001).

Landscapes are not only a geographical concept, but also a process of identity construction (Greider & Garkovich, 1994; Saugeres, 2002). Land has been used to maintain power among white men throughout Iowa's history. The first official European settlement in 1833 followed the Black Hawk Purchase. Not a 'purchase' at all, the federal government required the Sauk and Meskwaki to relinquish a portion of their land in eastern Iowa as punishment for resisting the federal government's 1829 requirement that they vacate their land in western Illinois (Harlan, 1931, p. 70). The Sauk and Meskwaki did eventually comply, but only after significant loss of life via the Black Hawk War. Soon thereafter, farms were established, and prairies tilled. As thousands of settler-colonizers poured into the state, railroads were constructed, and farms began to industrialize to meet demands for growing regional agricultural economies.

Farming remains central to Iowa's cultural identity. Understanding landscapes as symbolic constructs in addition to physical and geographic spaces helps us to understand the reflexivity between land use and identity. Proposed environmental changes may cause conflict when different groups hold conflicting meanings about land. Greider and Garkovich (1994) theorized that those in power impose a specific definition of the physical environment which reflects a group's symbols and meanings and it is easy to see how Iowa's physical landscape reflects the prioritization of agricultural productivity. Today, over 90 per cent of Iowa's land is used for agricultural production, and Iowa ranks first in the United States for production of corn, soybeans, hogs, and eggs (IDALS, 2014). Additionally, Iowa produces nearly 30 per cent of the nation's ethanol and exported over $11.3 billion in agricultural products in 2012 (IDALS, 2014). These productivist narratives are in opposition to narratives that centre the commons, or resources shared or held and protected in common, such as land, water, public health, and the rights of future generations.

The recent framing of water pollution in Iowa as a social problem has gained new importance, even though it has long been known that the state is one of the top contributors to the Gulf of Mexico's hypoxic zone (Alexander, Smith, Schwarz, Boyer, Nolan, & Brakebill, 2008). Today, farmland in Iowa is literally washing away through soil erosion, and the state's waterways are increasingly polluted by agricultural run-off (Cox, Hug, & Bruzelius, 2011; Naidenko, Cox, & Bruzelius, 2012). Given that only 1 per cent of Iowa's land is publicly-owned by state or federal government (NRC, 2010), the decisions of Iowa's farmland owners influence water and soil health downstream.

The saga of agricultural water pollution and soil loss in Iowa is an ongoing one, and one that positions rural Iowans against urban Iowans and farmers against environmentalists. Many in the state argue that voluntary conservation efforts are insufficient to address Iowa's agricultural pollution problem – a problem that has increasingly becoming an urban problem. The CEO of Des Moines Water Works, Bill Stowe, became an outspoken advocate for clean water, citing a steady increase in nitrate pollution in the water supply and increased financial burden to customers in the state's capital city, Des Moines. After several years of threatening lawsuits, the Des Moines Board of Water Works Trustees, representatives for the water utility serving the capital city's metro area of approximately 500,000, filed a federal Clean Water Act lawsuit against the supervisors of three agricultural tile drainage districts in counties northwest of Des Moines on 16 March 2015 (Tidgren, 2015). The case inspired a heated debate about responsibility for collective resources in which the agricultural industry threatened boycotts of Des Moines and urban residents villainized farmers; the case was ultimately dismissed in March 2017 (Eller, 2017). The proposed Dakota Access pipeline, however, offered an opportunity to shift the symbolic landscape of land and power in Iowa.

Iowa's statewide newspaper, the *Des Moines Register*, first broke the story about Dakota Access' proposed plan on 10 July 2014 (Petroski, 2014). The emergence of the Bakken Pipeline Resistance Coalition and their framing of the pipeline as a threat to the state's water, soil, and public health are situated within this ongoing statewide debate about water quality and agricultural pollution that was already well underway in Iowa. We were interested in the revision of cultural narratives about land in relation to this new contentious environmental problem – the Dakota Access pipeline – especially given the simultaneous debate in the state over the responsibility for clean water and the right to farm, as illustrated through the aforementioned Des Moines Water Works lawsuit. In order to better understand the Coalition's interaction with existing productivist agricultural narratives, we asked three questions: 1) How do Coalition participants articulate their participation in the resistance?, 2) How do their narratives compare to the narratives of agricultural productivity that dominate land use in Iowa?, and 3) What opportunities might these narrative shifts create to centre the commons?

Methodology

The pipeline resistance varied along the pipeline route from Standing Rock, North Dakota to Patoka, Illinois. We focus here on the framing of the pipeline resistance through the emergence of the Coalition in Iowa in 2014 and subsequent evolution in narratives up to the beginning of the pipeline's construction in Iowa in the summer of 2016. In keeping with constructivist grounded theoretical approaches, 'we can claim only to have interpreted a reality, as we understood both our own experience and our subjects' portrayals of theirs' (Charmaz, 2003, p. 271).

We, as active participant observers, acknowledge that our identities as Iowans and our various levels of involvement within the Coalition shaped the narratives to which we had access. Our fieldwork was shaped by our identities as white women descended from settler society. We do not claim that our experiences organizing resistance to the Dakota Access pipeline in Iowa represent everyone's experience organizing in Iowa or elsewhere, but believe this research on narrative reconstruction in Iowa – where settler culture and the elevation of productivist agricultural narratives have so influenced formation of power and identity in relationship to the land – might offer insights into shifts elsewhere.

We drew upon the recommendations of Buch and Staller (2007) in our research, maintaining transparency about our evolving roles along the spectrum of observer-participant. Our active participation in the resistance provided us access to many hours of observation and discussion during the formation of the Coalition and this informs our focus on the construction of resistance narratives. As researchers, we have engaged in processes of ongoing content analysis, observation, note-taking, reflection, comparison, and re-evaluation together since summer 2014. As participants, we first engaged in the resistance as community members – the pipeline's path cuts right through the county in which we both lived in central Iowa and where we were both, in 2014, graduate students of sociology and sustainable agriculture. We have each served on the board of Women, Food and Agriculture Network and been active members in Iowa State University's Sustainable Agriculture Student Association, two of the Coalition's founding members. Our own experiences, then, were 'starting points' for inquiry in this research (Devault, 1996; Smith, 1987).

We followed Loseke's (2012) recommendations for narrative analysis: first establishing the context of the existing dominant narratives in Iowa about land and its use; then engaging in close reading of ethnographic data from participant observation at public events and organizing meetings, the Coalition's statements, and individual Coalition members' testimonies in which we categorized explicit examples of narrative shifts and analysed symbolic codes; and, finally, using grounded theory to operationalize changes in how the resistance framed their opposition and collaboration. A social constructionist approach to grounded theory allowed us to analyse the emergence of resistance narratives, the process of resistance narrative construction, and how or why these narratives are reconstructed (Charmaz, 2003). As active participants, we had access to both the explicit narratives shared in public testimony, interviews, and at rallies, as well as informal conversations among claims-makers who strategized together to construct an emphasis on public goods and the commons.

We first share the story of the emergence of the Coalition as pipeline opponents began their collaborative resistance. We then identify how the Coalition, as a narrative-shifting arena, created space for a reframing of environmental conflicts through three strategic processes of narrative reconstruction: creating new space for claims-makers, elevating the commons, and aligning values beyond local conflict.

Emergence of the coalition and narrative shifts

Our findings suggest that the Coalition succeeded in creating a new narrative arena, a site in which reflexive revision of existing narratives about land and power are called into question, revised, and new constructions of narratives are sustained (Gubrium, 2005). In this new narrative arena, community members as well as farmland owners have authority to make claims about the future of the land and the land's value is defined through more collective means than yield per acre. We will first share the story of the Coalition's formation, followed by analysis of the narrative shifts made possible through the Coalition's reframing processes.

Emergence of the coalition

The narrative work sustained through the Coalition's organizing is, to date, ongoing even as oil flows through the pipeline. We focus here on the emergence of the Coalition because of its importance in setting the foundation for later collective action and continued engagement throughout the pipeline resistance.

Plans for the proposed Dakota Access pipeline – later to be known as the Bakken Pipeline – were announced on the front page of the *Des Moines Register* on 10 July 2014 (Petroski, 2014). Representatives from Iowa Citizens for Community Improvement (ICCI), Food and Water Watch (F&WW), Iowa State University's Sustainable Agriculture Student Association (ISU SASA), and Women, Food and Agriculture Network (WFAN) launched a Facebook page entitled 'Iowa Bakken Oil Pipeline Resistance' in late July 2014 (later renamed Bakken Pipeline Resistance Coalition). Soon thereafter, on 9 August 2014, these same allies, in addition to the environmental and agricultural groups including Iowa Sierra Club and Iowa Farmers Union, first met together in Des Moines to plan collaborative resistance to the proposed pipeline.

Members of ICCI, F&WW, WFAN, Iowa Sierra Club, and ISU SASA continued to coordinate collaborative anti-pipeline efforts. Together, representatives from these groups launched an online CREDO petition demanding that the governor of Iowa, Terry Branstad, stop the proposal.[2] On 14 October 2014, F&WW sponsored an event in Des Moines in solidarity with the Global Frackdown campaign. Activists from the various groups held an action and press conference, and were joined by landowners along the pipeline route.

Throughout the fall of 2014, the group continued to raise visibility about the pipeline in local media, educate the public about the project, and began to organize resistance through meetings in communities, in-person strategy meetings, and weekly phone calls for community members along the pipeline route, including landowners. The state's permitting authority, the Iowa Utilities Board, held mandated meetings along the proposed pipeline's route in December 2014 to provide information about the proposed pipeline project that were well-attended with concerned and, sometimes, angry community members across the state (Tauscheck, 2014).

Additionally, in December 2014, ISU SASA students coordinated what they intended to be a community meeting for residents of Story County, Iowa, the mid-way point of the pipeline project in the centre of the state and future home to the pipeline's pumping station. The response was overwhelming; roughly 300 people attended, many of whom drove from across the state to attend. Promotion for the meeting attracted statewide attention, and it soon became clear that many wanted to contribute funds to organized efforts against the pipeline. Pipeline opponents had previously decided to do their work through their various organizational partners rather than to seek 501c3 status as a collaborative entity. Science & Environmental Health Network became the fiscal sponsor of the Coalition in December 2014 when the Story County meeting and subsequent activities necessitated shared funds.

After the meeting, Iowa State University graduate students developed a website, Twitter account, and logo for the Coalition, and the Bakken Pipeline Resistance Coalition officially launched in February 2015. At its launch, farmland owners, union members, students, and concerned community members all shared public statements about their individual concerns about the pipeline project. The Coalition's members were united in agreement that the pipeline must be stopped, but differed in organizing tactics, size and staffing, and prioritization of concerns.

Shifting narratives: creating new space for claims-makers, elevating the commons, and aligning values beyond local conflict

The ongoing environmental debate in Iowa about water quality and agricultural pollution extenuated an already-existing duality among urban and rural residents, farmers and non-farmers, as the framing of this problem fit within the traditional framing of identity in dominant productivist agricultural narratives – the male farmer as power holder and decision maker (Carter, 2017). The Dakota Access pipeline, however, presented a new and unfamiliar threat from out-of-state. As Ordner (2017) explains in the case of Keystone XL's resistance in Nebraska, these new energy sacrifice zones create space for unlikely allies to work collectively to protect the commons. Similar to the Keystone XL resistance in Nebraska, being targeted by an outside threat helped claims-makers shift narratives beyond the duality framing of the water pollution social problem. Just as one may feel the pressure to tell the 'right' story (Hopper, 2001), expectations of time and place are also important to being heard as a claims-maker (Polletta, Chen, Gardner, & Motes 2011, p. 118). Through participation in the Coalition, claims-makers co-created new versions of the 'right' story about Iowa's farmland as the Coalition acted as a narrative shifting arena, redefining expectations about narratives related to land and land use when confronted with an outside threat (Gubrium, 2005).

The Coalition offered a space in which new and differing connections to land and community were valued. The narratives of agricultural land use in Iowa, and the subsequent revisions of these narratives to create narratives of pipeline resistance, reflect important shifts in identity: 'Landscapes are the reflection of these cultural identities, which are about *us*, rather than the natural environment' (Greider & Garkovich, 1994, p. 2). Farmland owners and the general public had the opportunity, through participation in the Coalition, to affirm narratives prioritizing the commons over production, confirming Polletta et al.'s (2011) finding that narratives can be changed through opportunities to learn new conventions. In doing so, the Coalition prepared fertile ground for long term and future collaboration against the pipeline project.

We identify three strategic processes used in the Coalition's collaborative organizing that redefined and shifted the resistance narrative: creating new space for claims-makers, elevating the commons, and aligning values beyond local conflict. While we present these in the order in which they first emerged, we emphasize that each is iterative and part of the ongoing narrative-shifting process.

Creating new space for claims-makers

In Iowa's ongoing water quality debate, agricultural claims-makers defend their right to farm as a livelihood against environmentalists' claims about agricultural pollution (see Haider, n.d.). The power dynamics rooted in settler society in Iowa privilege the voices of farmers and farmland owners when it comes to questions of farmland and its use. Throughout the process of the pipeline's permitting and approval, farmland owners along the proposed pipeline path stressed dominant narrative concerns related to abuse of eminent domain and the pipeline's potential ill effects on farmland yield and profitability. That those landowners in the direct path of the pipeline were ready claims-makers in opposition to the pipeline is not a surprise; however, the common cause – opposition to the pipeline – unified these landowners with an unlikely and ahistorical group of allies including indigenous, faith-based, renewable energy, wildlife, conservation, property rights, environmental, agriculture, and social justice groups.

The Coalition's mission – 'We are a coalition of organizations representing landowners, community members, non-profits, and interest groups united to stop the construction of the Bakken Pipeline' – opened the resistance to concerned community members other than farmland owners, as well as to farmland owners with concerns beyond the familiar NIMBY debate about eminent domain and farmland. In doing so, participants in the resistance, both landowners and others, were able to co-construct narratives about their motivations and participation in the resistance in terms other than profitability and productivity, challenging and shifting the dominant narrative.

Student organizers of the December 2014 meeting in Ames, Iowa, the first large convening in Iowa of pipeline opponents, advertised with a poster referencing the danger of a crude oil pipeline in agricultural fields. This poster appealed to the dominant narrative, posing the pipeline as a threat to Iowa's commodity production. However, the organizers of the meeting purposefully assembled a panel of perspectives to share information about the economic, environmental, social justice, and legal risks posed by the pipeline. The poster and the early inclusion of landowners within the Coalition's events are examples of how the Coalition worked to use the dominant narrative as a starting point or entry way into larger discussions about the commons.

At the Coalition's formal launch in February of 2015, the Coalition organizers continued to emphasize the intentional coordination of new claims-makers from various interest groups, inviting those new to activism to apply their paradigms to the anti-pipeline struggle. Further, the Coalition was careful to elevate the stories of community members opposed to the pipeline – not only those stories of landowners' opposition. For example, their press release from the February 2015 launch included the following quote from Andrea Basche of the Iowa State University Sustainable Agriculture Student Association:

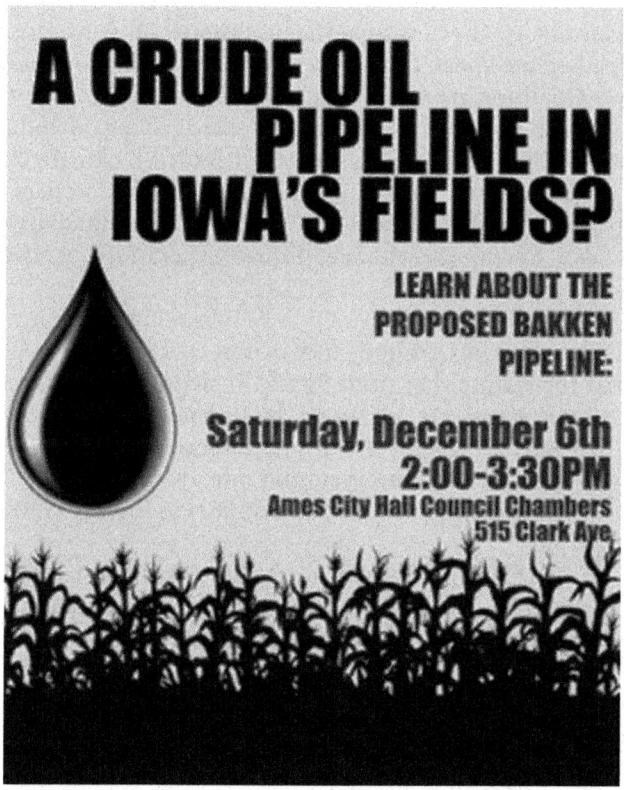

Figure 7.1 Bakken Pipeline Resistance Coalition meeting promotion poster (used with permission).

This pipeline is one more project that sets our energy future in the wrong direction and continues on a path that leaves future generations with a larger carbon debt. Where is the justice in that? We know the future we need is less carbon, so why wait to get started?

(BPRC, 2015a)

At the time of launch, members included the following 20 state and national non-profit, grassroots, student, and single-issue groups: 100 Grannies for a Livable Future, 1000 Friends of Iowa, Allamakee County Protectors, Citizens' Climate Lobby, Drake Environmental Action League, Food & Water Watch, Iowa Audubon Society, Iowa Citizens for Community Improvement, Iowa Climate Advocates, Iowa Farmers Union, Iowa Interfaith Power & Light, Iowa Pipeline Abatement Group, Iowa Renewable Energy Association, Iowa State University ActivUs, Iowa State University Sustainable Agriculture Student Association, No Bakken Here, Science & Environmental Health Network, Sierra Club (Iowa Chapter), Women, Food & Agriculture Network, and Women's International League of Peace and Freedom (BPRC, 2015a).

This intentionality in elevating the multitude of voices opposing the pipeline helped create space for people to challenge the dominant narrative in new ways. Coalition members were careful to also engage the broader community around issues of importance to their groups, including environmental sustainability, community, future generations, climate change, indigenous sovereignty, and corporate power. Soon, landowners along and near the pipeline route were speaking not only to productivity and profitability, but also to climate change, future generations, and community justice, among other concerns:

This pipeline in our ground and waterways would be a giant step backward for a state that prides itself in smart economic and environmental investments. As an Iowa landowner – but more so as an Iowan – I urge everyone to learn more about how this pipeline would affect our land, our communities, and our overall health.

(Kathy Holdefer, Jasper County landowner (BPRC, 2015a))

By the fall of 2015, the emphasis on landowners' protecting individual property had shifted to landowners' stewardship of the collective commons, as illustrated by a poster at a Coalition-sponsored rally before an Iowa Utilities Board hearing in Boone, Iowa (Figure 7.2).

This shift from protection of individual property to protection of the commons continued to evolve. Richard Lamb, a landowner whose multigenerational farm was threatened by eminent domain to construct the pipeline, was very vocal in his family's opposition not only to the pipeline crossing his farm but to its construction anywhere. A year after the Coalition's official launch, Lamb reported in an interview with ThinkProgress:

Figure 7.2 Poster at Bakken Pipeline Resistance Coalition rally, November 2015 (used with permission).

I've publicly and loudly state I would never, ever, willingly for any amount of money [allow] that pipeline cross my land. We don't think it's the right thing to be doing. We are against fracking, against fossil fuels, want to do what we can to avoid global warming, and this of course is contrary to all of that.

(Fragoso, 2016)

At a press conference on Richard Lamb's farm, Donnielle Wanatee, of Iowa's Sac and Fox Tribes/Meskwaki Nation, aligned landowners' rights with the collective commons of those living in the state: 'The Army Corps of Engineers needs to deny this permit because of the damage and threats to our ecosystem, our drinking water and our Iowa landowner's farmland' (Harrington, 2016). Pam Alexander, a landowner who attended the state's first action against the pipeline in October 2014 and has continued throughout the resistance to be a vocal opponent, shared her views on Iowa Public Radio:

I think it's going to be disastrous for our agricultural land. The state of Iowa's water is at risk, our land is at risk, and it's going to be a mess if this thing leaks. I would hate to be responsible for that.

(Moon & Kieffer, 2016)

In creating pathways for concerned groups and individuals to enter the resistance from various perspectives, the Coalition implicitly and explicitly challenged the dominant productivist narratives defining landscapes in terms of profit maximization and productivity. Though at first landowners were viewed as the only legitimate claims-makers, this shifted as Coalition members representing various interest groups advocated against the pipeline in terms of community, future generations, and environmental sustainability.

Elevating the commons

The synergy of these unlikely allies created new space for claims-makers to centre their shared opposition and, eventually, to elevate the commons. Consistent with narrative framing in the Keystone XL pipeline opposition in Nebraska (Ordner, 2016), the Coalition was successful in framing resistance around protection of the state's land, water, and common resources, or the commons.

Intentional framing of resistance by the coalition in terms of the commons took many forms. From press releases to protest signs and apparel, the coalition clearly emphasized and prominently featured the importance of common resources such as land, water, and future generations. For example, the Coalition chose blue for the colour of their t-shirts – distributed at events across the state – to represent water. The front of the shirt states 'Stop the Pipeline', echoing the common cause of the Coalition's mission statement. The back of the shirt emphasized three key points of the Coalition's reframing: 'Protect Iowa's Soil & Water, Landowners' Rights, Future Generations'.

In 2015, as the Iowa Utilities Board began to host public hearings during the start of the permitting process, Iowans emphasized their connection to Iowa's cultural identity as a farming state and the importance of Iowa's soil as a public good. Melonie Stall, a resident of a rural town near the pipeline's proposed path, testified that

> Iowa soil does not come with a price tag. It cannot be bought by a company that only stands for big profits. Members of the Iowa Utilities Board, I ask that you stand with me – with Iowa – and vote to oppose the Bakken Oil Pipeline.
>
> (BPRC, 2015c)

Dr. Tom Fenton, Professor Emeritus at Iowa State University, wrote a short summary of the risks the pipeline poses to Iowa's soil – from construction to eventual leakage – and the Coalition shared this resource with the public and policy makers alike (BPRC, 2015b). For those near the pipeline construction's disruption of soil and the threat of future leaks inspired many to document Dakota Access' work as 'watchdogs' and to submit construction violations to county and state authorities. In this way, the community elevated the shared responsibility to the state's soil health and the importance of soil as a public good. As Arlene Bates, a landowner in the pipeline's path asserted, 'Once you disturb the soil, it's gone' (DaSilva, 2016).

Emphasis on soil was joined by an emphasis on water. In early April 2016, LaDonna Brave Bull Allard and Joye Braun founded the Camp of Sacred Stones at Standing Rock in North Dakota in opposition to the proposed Dakota Access pipeline. The unifying message from Standing Rock – Water is Life! – helped to lift up the visibility of the pipeline's threat to all downstream. In the summer of 2016, the Coalition planned several flotillas along Iowa's rivers threatened by the pipeline's construction.

The first flotilla along the South Skunk River near Oskaloosa, Iowa was the idea of Sylvia (Rodgers) Spalding, a landowner currently managing her family's multi-generational farm from out-of-state. At the press conference prior to the flotilla's launch, Sylvia stood by Donnielle Wanatee from Iowa's Sac and Fox Tribes/Meskwaki Nation. Both spoke of the importance of stopping the pipeline, their shared heritage on the land, and the future of the river's health. Later, Sylvia wrote an essay about her opposition to the pipeline that was published in several state papers in slightly shorter form as an editorial. In the essay, Sylvia reflected upon that day along the South Skunk River and what she saw as her responsibility to the land:

> A great law of the Iroquois Confederacy is to live and work for the benefit of the seventh generation into the future. I am the 7th generation to descend from the white settlers on this Iowa land along the South Skunk River, and I am working now to protect not only that land and the river for my daughter and the six generations after her but also the many other lands and waters that are threatened by the fracking, transporting and burning of oil. My nightmare is the pipeline bursting and crude oil engulfing the floodplain forest of our family land. But I also fear for those already being impacted by climate change and what millions of barrels of additional oil and their greenhouse gasses will do to those I have met who inhabit the coastlines of Alaska, the Pacific Islands, the Pacific Northwest and the mouth of the Mississippi River; to the crop lands on low lying islands being encroached by saltwater intrusion; to the traditional salt gathering sites in Hawaii being inundated; the coral reefs that are bleaching; the native communities that are already having to relocate and migrate and the social and cultural upheaval this is causing.
>
> (Spalding, 2016)

This focus on protection of the commons was again the focus at a later flotilla along the Des Moines River in central Iowa. Mark Edwards, a community member, had spent many hours that summer meeting with landowners along the pipeline's proposed Des Moines River crossing, building trust and learning about their concerns. Surrounded by members from the Meskwaki tribe, landowners, environmentalists, and recreation lovers, the Coalition held a press conference along the Des Moines River prior to launching the flotilla and shared, again, the responsibility to protect the commons:

> We have a duty, a right to protect the public trust we share in common – the climate, the air, the land, this river. People have been living here for thousands of years without the threat of losing their water. One spill for one hour could dump one million gallons of oil into this river. Are we really willing to risk this? Are we really willing to allow it?
>
> (Mark Edwards, Boone County resident, speaking along the Des Moines River, 25 July 2016 (BPRC, 2016))

This elevation of the commons was a new focus for narratives about land and land use in Iowa, especially in connection to water as a public good. In the midst of Iowa's water quality controversy and its polarizing narratives, it was an unusual scenario to have landowners and environmentalists standing together, sharing their concerns about the commons. In the midst of a heated battle over water as a commodified resource, the Coalition was able to redefine water and land as *common resources*. Individuals and organizations used this narrative to garner support for resistance to the pipeline. Ultimately, collaborations among farmers, environmentalists, and other interest groups who had not previously collaborated for unified cause now have a shared narrative. These alliances positioned the Coalition well to connect to resistance beyond Iowa.

Aligning values beyond local conflict

The public framing of the pipeline debate in Iowa shifted over time through the Coalition's narrative work. Though the existing dominant narrative focused on the land's ability to produce maximum yield for maximum profit, the Coalition was able to create new space for claims-makers to elevate the commons and, eventually, to align with regional conflicts by lifting up shared Iowa values. These values include neighbouring and the importance of legacy through the protection of future generations' water, soil, and health. These shared values were key to connecting the Iowa pipeline struggle to the growing resistance at Standing Rock as well as to regional and national environmental and climate justice campaigns. As the movement to stop the Dakota Access pipeline grew, the Coalition began to work more closely with regional allies, including Minnesota 350. org, residents in southwestern Illinois, and, eventually, the resistance camps at Standing Rock.

Early on, Coalition members reached out to and collaborated with Keystone XL pipeline fighters in Nebraska to learn more about organizing grassroots resistance to pipeline construction. This enabled the Coalition to share resources and learn tactics, but also helped lift up regional struggles and broader issues of climate and social justice from the very beginning. The late Rev. Dr. Barbara Schlachter of Coalition partner 100Grannies for a Liveable Future and Iowa City Climate Advocates emphasized the connection between Iowa's pipeline fight and larger climate justice movements in May of 2015:

> We cannot allow 570,000 barrels of volatile crude to run through our state every day. While the pipeline has the potential to damage some of the world's finest farm land, it goes into the collective atmosphere, and we all experience the fallout of global warming, whether we live in Muscatine [Iowa] or Memphis or Moscow or Mumbai.
>
> (Schlachter, 2015)

In their alliance with those at Standing Rock and along the pipeline route, and with climate justice actions beyond Iowa, the Coalition members emphasized the importance of being a good neighbour. As Huxley, Iowa community member Brenda Brink stated at a rally outside the Iowa Utility Board office in October 2015, Iowans felt they had an obligation to a larger justice fight beyond the state's borders and were inspired to step up and play their part:

> I'm here [at this rally] because we need to stop this right now. There's a precedent getting ready to be set in Iowa. It's in my backyard and I don't want to see it anymore. We had the farmers, we had people that I hadn't seen at a rally before, we had people who were talking about climate justice from their own backyards where the corporations have come in and they've taken control of government officials through bribery … That's what we're fighting against, we're fighting against that big oligarchy and lack of social justice. We're just one little part of it, but it's an important part.
>
> (Harrington, 2015)

In an effort to build alliances beyond Iowa, coalition partners communicated via phone, list-servers, and in-person with various interest groups resisting the Dakota Access pipeline beyond the state, and with interest groups resisting other oil infrastructure projects. This collaboration was evident at various events hosted by the Coalition. Oftentimes, people resisting the pipeline outside of Iowa attended events in Iowa. For example, Joye Braun, who would later co-found Standing Rocks' Camp of Sacred Stones, travelled to Iowa for the Coalition's rally before the Iowa Utilities Board public hearing for the pipeline in Boone, Iowa on 13 November 2015 with her husband, Floyd Braun. Joye was joined by others from outside of Iowa, including climate justice activists from Southern Illinois, in providing public testimony against the pipeline. Later, in April of 2016, Joye and Floyd returned to conduct non-violent direct-action trainings in Des Moines, Iowa with Coalition members.

Similarly, Iowans involved in the Coalition attended events outside the state of Iowa. The founding of the Camp of Sacred Stones in early April 2016 elevated the pipeline resistance into a global campaign and the Coalition supported the camp's ongoing efforts. Coalition partner Science & Environmental Health's executive director, Carolyn Raffensperger, provided legal support to the early organizers of the resistance camps at Standing Rock. Many from the Coalition visited Standing Rock – some for short supply runs, others for longer visits or multiple visits. Additionally, members from the Coalition attended hearings at Standing Rock to bear witness to the tribes' testimony, and the Coalition regularly sent up financial and supply donations throughout the camps' existence. As Iowan Peter Clay, a member of Coalition partner Citizens Climate Lobby, explained:

> My car was filled with non-perishable food, a new chain saw (with extra chain and related supplies) as well as a wood maul and sundry other items that had been asked for. The BPRC, along with some of my friends, donated all of this. Driving for the first time across the vast and dramatic landscapes of what white people (wašíču in the Lakota language) call North and South Dakota, I was struck by the beauty of the land and haunted by what I know of its history. [...] Listening respectfully to the Lakota people living at or visiting the Spirit Camp each day gave me a new respect for the First Nations of North America and for their values, cultures and deep reverence for the earth. I was changed and challenged in many ways and I continue to reflect upon my experience. It was a privilege to represent Iowans standing in solidarity with the courageous Northern Plains tribes against the pipeline. My plan is to return again. Other Iowans are encouraged to visit too. We are much stronger together!
>
> (Clay, 2016)

In the summer of 2016, pipeline construction began in Iowa. Using lessons learned at Standing Rock, water protection camps sprang up in Iowa along the Des Moines River near Pilot Mound, Iowa and the Mississippi Stand camp in southeast Iowa along the Mississippi River. In March of 2017, Coalition partner Indigenous Iowa founded Little Creek Camp near Williamsburg, Iowa. Little Creek Camp's purpose was to bring together allies working towards indigenous sovereignty and decolonization. Oil began flowing through the pipeline on 1 June 2017, yet Little Creek Camp continued through the fall and the Coalition continued to organize in collaboration with local and national groups. This continued alignment with local, regional, and national extractive energy resistance and climate justice campaigns inspires new meaning to the words 'neighbour' and 'home', creating space for Coalition partners to continue to engage in ongoing efforts to shut down extractive energy projects and plan for a longer time frame than one pipeline fight.

Conclusion

Emerging local, regional, and national resistance to extractive energy projects, including the Dakota Access pipeline, provide us real time examples of the work narratives do in reclaiming and reconstructing the commons. The Dakota Access pipeline presented a new player on Iowa's agricultural landscape – an out of state oil company – whose threat unified a new front of extractive energy opponents in Iowa. Whatever their original motivation for working together, the many members of the Bakken Pipeline Resistance Coalition welcomed all to do their part to protect what was at stake. The Coalition shifted power from historically legitimate claims-makers about rural spaces (farmland owners and farmers) to opponents of the pipeline as a collective. Rather than focus solely on farmland owners' private property rights, opponents – including farmland owners – incorporated a broader reaching narrative emphasizing water, soil, and public health as the commons. The story of the Bakken Pipeline Resistance Coalition's narrative work is one story of resistance to the Dakota Access pipeline, and one story of the many localized grassroots campaigns against extractive energy across the globe.

The Coalition successfully created a space in which a shared cause motivated the building of a new grassroots power, and the work continues on. Since the summer of 2014, the Bakken Pipeline Resistance Coalition has tended to and cared for a shifting symbolic landscape – one that prioritizes the commons and in which people from rural and urban spaces, indigenous communities and settler society, recreational water lovers, climate justice fighters, and renewable energy advocates all have a voice. As we write this, oil flows beneath Lake Oahe at Standing Rock in North Dakota and under the Mississippi River near Keokuk, Iowa. Still, the Coalition continues its narrative work, organizing legal appeals and community events in effort to hold the state and Dakota Access accountable for what they see as a trespass of both private land and public good. 'We cannot live without stories, big stories finally, to tell us what is real and significant, to know who we are, where we are, what we are doing, and why', writes Christian Smith (2003, p. 67). In the Coalition's resistance narratives, we find a powerful and far-reaching story far less new than ancient – a story prioritizing protection, neighbouring, and home.

Acknowledgements

We thank members of the Bakken Pipeline Resistance Coalition and water protectors everywhere.

Notes

1 The Dakota Access pipeline – a crude oil pipeline moving hydraulically fractured oil from the Bakken fields of North Dakota through North Dakota, South Dakota, Iowa, and into Illinois – was originally referred to as the 'Bakken pipeline' by pipeline opponents and proponents in Iowa.
2 The original CREDO petition can be retrieved here: www.credomobilize.com/petitions/tell-governor-branstad-stop-the-iowa-bakken-oil-pipeline.

References

Alexander, R. B., Smith, R. A. Schwarz, G. E., Boyer, E. W., Nolan, J. V., & Brakebill, J. W. (2008). Differences in phosphorus and nitrogen delivery to the Gulf of Mexico from the Mississippi river basin. *Environmental Science & Technology*, 42, 822–830.
BPRC – Bakken Pipeline Resistance Coalition. (2015a). Launching the Coalition. 2 February 2015. Retrieved 26 March 2017 from http://nobakken.com/2015/02/02/launching-the-coalition/.
BPRC – Bakken Pipeline Resistance Coalition. (2015b). Bakken Pipeline May Damage Soil Conditions for Generations. 10 October 2015. Retrieved 26 March 2017 from http://nobakken.com/2015/10/06/bakken-pipeline-may-damage-soil-conditions-for-generations/.
BPRC – Bakken Pipeline Resistance Coalition. (2015c). IUB Public Hearing a Win for Farmers, Landowners, and Community Members. 13 November 2015. Retrieved 26 March 2017 from http://nobakken.com/2015/11/13/iub-public-hearing-a-win-for-farmers-landowners-and-community-members/.
BPRC – Bakken Pipeline Resistance Coalition. (2016). We Protect Our Rivers! 27 June 2016. Retrieved 26 March 2017 from http://nobakken.com/2016/06/27/we-protect-our-rivers/.
Berger, P. L. & Luckmann, T. (1967). *The Social Construction of Reality: A Treatise in the Sociology of Knowledge*. Garden City, NY: Anchor Books, Doubleday & Company.
Buch, E. D. & Staller, K. M. (2007). The feminist practice of ethnography. In S. N. Hesse-Biber & P. L. Leavy (Eds), *Feminist Research Practice: A Primer* (pp. 187–221), Thousand Oaks, CA: SAGE Publications.
Carter, A. (2017). Placeholders and changemakers: Women farmland owners navigating gendered expectations. *Rural Sociology*, 82, 3, 499–523.
Charmaz, K. (2003). Grounded theory: Objectivist and constructivist methods. In N. K. Denzin & Y. S. Lincoln (Eds), *Strategies of Qualitative Inquiry* (pp. 249–291) Thousand Oaks, CA: SAGE Publications.
Clay, P. (2016). Voices Against the Pipeline: BPRC Supports Spirit Camp in North Dakota. 21 June 2016. Bakken Pipeline Resistance Coalition. Retrieved 1 April 2017 from http://nobakken.com/2016/06/21/voices-against-the-pipeline-bprc-supports-spirit-camp-in-north-dakota/.
Cox, C., Hug, A. & Bruzelius, N. (2011). Losing Ground. Environmental Working Group. Retrieved 20 April 2012 from http://static.ewg.org/reports/2010/losingground/pdf/losingground_report.pdf.
DaSilva, M. (2016). Two Iowa Landowners Vow to Continue Fight against Bakken Pipeline. 6 June 2016, updated 7 June 2016. Retrieved 26 March 2017 from

http://whotv.com/2016/06/06/two-iowa-landowners-vow-to-continue-fight-against-bakken-pipeline/.

Devault, M. L. (1996). Talking back to sociology: Distinctive contributions of feminist methodology. *Annual Review of Sociology*, 22, 29–50.

Eller, D. (2017). With Water Works' lawsuit dismissed, water quality is the legislature's problem. *Des Moines Register*. n.p., 17 March 2017. Retrieved 19 March 2017 from www.desmoinesregister.com/story/money/agriculture/2017/03/17/judge-dismisses-water-works-nitrates-lawsuit/99327928/.

Federici, S. & Caffentziz, G. (2013). Commons against and beyond capitalism, *Upping the Anti: A Journal of Theory and Action*, 15 September, 83–97.

Fragoso, A. D. (2016). Iowa Will Soon Decide Whether to Allow an Oil Company to Seize Residents' Land. 9 February 2016. ThinkProgress. Retrieved 2 April 2017 from https://thinkprogress.org/iowa-will-soon-decide-whether-to-allow-an-oil-company-to-seize-residents-land-ed87580719a6.

Greider, T. and Garkovich, L. (1994). Landscapes: The social construction of nature and the environment. *Rural Sociology*, 59, 1, 1–24.

Gubrium, J. F. (2005). Introduction: Narrative environments and social problems. *Social Problems*, 52, 4, 525–528.

Haider, M. (n.d.). Farmers Clash With Des Moines Water Works Over Lawsuit. Nexstar Broadcasting. Retrieved 28 March 2017 from www.siouxlandproud.com/news/local-news/farmers-clash-with-des-moines-water-works-over-lawsuit.

Harlan, E. R. (1931). *A Narrative History of the People of Iowa*. Chicago, IL: American Historical Society.

Harrington, A. (2015). Protesters flood IUB with objections over Bakken Pipeline. *Ames Tribune*. Retrieved 1 April 2017 from www.amestrib.com/news/protesters-flood-iub-objections-over-bakken-pipeline.

Harrington, A. (2016). Landowners rally to challenge eminent domain, Bakken pipeline. *Boone News Republican*. Retrieved 1 April 2017 from www.news-republican.com/news/local/landowners-rally-challenge-eminent-domain-bakken-pipeline.html.

Hopper, J. (2001). Contested selves in divorce proceedings. In J. F. Gubrium & J. A. Holstein (Eds), *Institutional Selves: Troubled Selves in a Postmodern World* (pp. 127–141). New York: Oxford University Press.

IDALS – Iowa Department of Agriculture and Land Stewardship. (2014). *Iowa Agriculture Quick Facts: Quick Stats*. (Revised July 2014). Retrieved 10 June 2015. From www.iowaagriculture.gov/quickfacts.asp.

Joubert, B. & Davidson, D. J. (2010). Mediating constructivism, nature and dissonant land use values: The case of northwest Saskatchewan Métis. *Human Ecology Review*, 17, 1, 1–10.

Loseke, D. (2007). The study of identity as cultural, institutional, organizational, and personal narratives: Theoretical and empirical integrations. *The Sociological Quarterly*, 48, 4, 661–688.

Loseke, D. (2012). The empirical analysis of formula stories. in J. A. Holstein & J. F. Gubrium (Eds), *Varieties of Narrative Analysis* (pp. 251–272). Thousand Oaks, CA: SAGE Publications.

McAdam, D. & Boudet, H. S. (2012). *Putting Social Movements in their Place: Explaining Opposition to Energy Projects in the United States, 2000–2005*. New York: Cambridge University Press.

Moon, L. & Kieffer, B. (2016). After Eminent Domain Rights, Landowners Continue to Fight Bakken Pipeline. Iowa Public Radio. Retrieved 24 March 2017 from http://iowapublicradio.org/post/after-eminent-domain-rights-granted-landowners-continue-fight-bakken-pipeline#stream/0.

Naidenko, O. V., Cox, C., & Bruzelius, N. (2012). Troubled Waters: Farm Pollution Threatens Drinking Water. Environmental Working Group. Retrieved 20 April 2012 from http://static.ewg.org/reports/2012/troubled_waters/troubled_waters.pdf.

NRC – National Resources Council of Maine. (2010). Public Land Ownership by State. Retrieved 10 June 2015 from www.nrcm.org/documents/publiclandownership.pdf.

Ordner, J. P. (2016). *Protecting the Good Life: Grassroots Resistance to the Keystone XL Pipeline in Nebraska*. (Unpublished doctoral dissertation) Lawrence, KS: University of Kansas.

Ordner, J. P. (2017). Commentary: Community Action and Climate Change. *Nature Climate Change*, 7, 161–163.

Petroski, W. (2014). Oil Pipeline Across Iowa Proposed. *Des Moines Register*. 10 July 2014. Retrieved 28 March 2017 from www.desmoinesregister.com/story/news/2014/07/10/oil-pipeline-across-iowa-proposed/12445185/.

Polletta, F., Chen, P. C. B., Gardner, B. G., & Motes, A. (2011). The sociology of storytelling. *Annual Review of Sociology*, 37, 109–130.

Saugeres, L. (2002). The cultural representation of the farming landscape: Masculinity, power and nature. *Journal of Rural Studies*, 18, 4, 373–384.

Schlachter, B. (2015). Voices Against the Pipeline: The Rev. Dr. Barbara Schlachter. Bakken Pipeline Resistance Coalition. 6 May 2015. Retrieved 2 April 2017 from http://nobakken.com/2015/05/06/voices-against-the-pipeline-the-rev-dr-barbara-schlachter/.

Smith, C. (2003). *Moral, Believing Animals: Human Personhood and Culture*. New York: Oxford University Press.

Smith, D. (1987). *The Everyday World as Problematic: A Feminist Sociology*. Boston, MA: Northeastern University Press.

Spalding, S. R. (2016). Voices Against the Pipeline: 'Why the River, Why Standing Rock'. Bakken Pipeline Resistance Coalition. 9 September 2016. Retrieved 26 March 2017 from http://nobakken.com/2016/09/26/voices-against-the-pipeline-why-the-river-why-standing-rock-by-sylvia-spalding/.

Tauscheck, M. (2014). 300 Attended Ankeny Meeting on Proposed Pipeline. KCCI news. Retrieved 28 March 2017 from www.kcci.com/article/300-attended-ankeny-meeting-on-proposed-pipeline/6434865.

Tidgren, K. A. (2015). Des Moines Board of Water Works Trustees Files Lawsuit. March 16, 2015. Iowa State University Center for Agricultural Law and Taxation (CALT). Retrieved 10 June 2015 from www.calt.iastate.edu/article/des-moines-board-water-works-trustees-files-lawsuit.

Trauger, A. (2001). Women farmers in Minnesota and the post-productivist tradition. *The Great Lakes Geographer*, 8, 2, 53–66.

Vanderpol, L. (2016). Landowners say pipeline 'not a done deal'. *The Oskaloosa Herald*. 24 July 2016. Retrieved 26 March 2017 from www.oskaloosa.com/news/local_news/landowners-say-pipeline-not-a-done-deal/article_134ce9e8-51f8-11e6-8e10-377c7e47e074.html.

Zibechi, R. (2012). *Territories in Resistance: A Cartography of Latin American Social Movements*. Oakland, CA: AK Press.

8 Discursive framing and community mobilization

Stopping the Melancthon Mega Quarry in Ontario, Canada

Rebecca McEvoy and John F. Devlin

Introduction

The preservation of agricultural land has resulted in many contentious development proposals for land use near urban areas (Gayler, 1982). Melancthon Township located about 100 kilometres northwest of the City of Toronto is primarily agricultural land and has one of the lowest population densities in southwestern Ontario, Canada. In April 2011, an application for a licence to quarry limestone was submitted to the Ontario Ministry of Natural Resources (MNR) by The Highland Companies, a Nova Scotia-registered company backed by Baupost, a Boston-based hedge fund. The application proposed the development of a 2,316-acre open-pit quarry for mining amabel dolostone bedrock (Shuff, 2011). The application gave rise to a successful movement to stop the mega-quarry. Over one and a half years, the movement built complex alliances and organized unique events linking the farming community around Melancthon to food and environmental movements that were regional and national in scope.

The social movement engaged the local farmers and property owners around the proposed quarry site and interested individuals who lived within the potentially affected watershed and foodshed including the Toronto region. The diversity of movement adherents was notable. The David Suzuki Foundation, a national environmental advocacy organization, was approached to solicit their expertise and tap into their network of environmental activists. The Foundation used its contact list of subscribers across Ontario to spread the word. The Foundation also connected the residents of Melancthon to a broader national network of environmental activists. Another example of alliance building was the work of Artists against the Mega Quarry, a group that held art shows and sales with landscapes of the Melancthon countryside being auctioned off to raise funds. On 21 November 2012 the application for the extraction licence was withdrawn citing community opposition as a major reason. Social movement opposition efforts that successfully stop development projects are rare (Devlin, Yap, & Weir, 2005; Devlin, 2006; Devlin & Yap, 2008). This chapter outlines the genesis, process and outcomes of the Stop the Mega-Quarry movement. It demonstrates how creative framing focusing on food allowed the movement to gain widespread support and highlights the important role of technical allies in supporting the movement's environmental critique. The chapter is based on print and digital documents including newspaper articles, blog posts, government documents, academic studies[1] and interviews with key informants centrally engaged in the movement.

Emergence: a movement is born

In 2006, The Highland Companies began purchasing farmland in Melancthon Township. By 2011 they owned roughly 3,400 hectares of prime agricultural land including two of the most significant potato farming operations in the township and had become the largest grower and distributor of potatoes in the Province of Ontario (The Highland Companies, 2012, as cited in Bell, Isaac, Jamal, Levay, & Wright, 2012, p. 4).

Local landowners, farmers, and concerned citizens began to suspect that the Highland Companies had plans beyond farming. Unconventional activities had been seen taking place on the property such as well testing and drilling, archaeological studies, and the demolition of farm buildings (Waters, 2011b). The Highland Companies insisted that they were good farmers intent on growing potatoes, but many neighbours were concerned. In January 2009, at the arena in Honeywood, the first public meeting was held to discuss the intentions of this new landowner. To the organizers' delight people were lined up outside and waiting to get in (Interview, 15 March 2016).

At this meeting the Highland Companies representative denied that there were plans to build a mega quarry. But one community member had already called the Ministry of Natural Resources (MNR) asking about the company and had been told that the company had been talking to the MNR for several years about a massive quarry application (Interview, 15 March 2016).

When it was clear that The Highland Companies intended to develop a quarry, the meeting moved to form the North Dufferin Agricultural and Community Taskforce (NDACT). This group would become the most prominent opponent of the proposed mega quarry (Bell et al., 2012, p. 22). NDACT became an incorporated, not-for-profit entity whose executive board was comprised of volunteers from Mulmur and Melancthon, two townships near the proposed quarry site in Dufferin County (*NDACT History*, n.d.). NDACT's mission would be to

> preserve and protect the unique and non-renewable resources of North Dufferin County – including the Headwaters that supply Water to hundreds of thousands of Ontarians', our exceptional Prime Agricultural Farmland and the environment, social economic and cultural

characteristics that have been such an important and vibrant part of our community and its heritage for more than 150 years.

(NDACT's Mission Statement, n.d.)

Within a few days of the initial community meeting, The Highland Companies stated that they had never denied their intention to quarry the land. But two more years passed before the company made a formal application. In April 2011 the Highland Companies submitted its application to the MNR for a licence to extract aggregate from the land they had amassed in Melancthon. Receipt of a Class A licence under Section 7(2)(a) of the Aggregate Resources Act (ARA) would allow them to remove more than 20,000 tonnes of aggregate annually from an open-pit limestone quarry for amabel dolostone used as building stone or crushed stone. The proposed quarry would cover 937 hectares of prime agricultural land, which is about one-third the size of downtown Toronto (Bell et al., 2012, p. 4).

The MNR defines a 'mega-quarry' as any quarry in which the total rock reserve exceeds 150 million tonnes. The application submitted by The Highland Companies proposed the development of a quarry for which the rock reserve was 1 billion tonnes (Shuff, 2011). The proposed quarry would reach a depth of 200 feet below the water table. By comparison Niagara Falls has a depth of just 167 feet. The development would impact two major watersheds and require that 600 million litres of water be pumped each day in perpetuity to prevent the pit from flooding (Shuff, 2011). The plan also included a proposal to build a rail line from the quarry to Owen Sound which would connect the quarry to the Great Lakes and facilitate movement of aggregate into the United States (Keller, 2011).

At the time the policy context in rural Ontario was highly favourable to the aggregate extraction industry. Aggregate, which includes sand, gravel, clay, shale, stone, limestone, dolostone, sandstone, marble, and granite (Binstock & Carter-Whitney, 2011), was in high demand for construction projects. The Greater Toronto Area was the fastest growing region in Ontario with a population that was anticipated to increase by almost three million by 2041 (Ministry of Finance, n.d.). These projections, coupled with the reality that aggregate is heavy and therefor expensive to transport, meant that obtaining access to aggregate deposits close to areas of growth was a priority for the Province. A study on the state of aggregate resource in Ontario acknowledged that high quality aggregate deposits located near the Greater Toronto Area (GTA) were dwindling (Ministry of Natural Resources, 2010). If new licences were not approved there would be shortages. This would increase the cost of aggregates as the supply would have to be sourced from other parts of the province (Binstock & Carter-Whitney, 2011).

In addition, quarries in Ontario do not require an environmental assessment. Applicants for a licence are required to submit an application, site plan, and technical reports to the Ministry of Natural Resources (MNR).

But policy analysts asserted that enforcement of the Aggregate Resources Act (ARA) resulted in inadequate environmental monitoring; ineffective inspection procedures; and an absence of rehabilitation measures leading to undue environmental management, health, and safety risks (Binstock & Carter-Whitney, 2011).

The quarry proposal did outline the corporation's plan to rehabilitate the land at the bottom of the quarry for agricultural use. But the plan for rehabilitation below the water table would require continuous pumping of 600 million litres of water each day (Stadtländer, 2011). The potential life-span of the operation would be 50 to 100 years (Bell et al., 2012). The proposal also stated that the land owned by The Highland Companies which was not included in the quarry footprint would continue to be farmed (Bell et al., 2012, p. 4).

Highland's claims

The Highland Companies had been vague or deliberately deceptive about plans for the quarry development. This began as early as the farmland acquisition in 2004. John Lowndes, a local man who approached the farmers on behalf of The Highland Companies, suggested the company wanted to be the largest potato farming operation in the province but made no mention of their intention to develop a quarry. As such, the farm-land was sold under false premises. In a 2011 interview with CBC News, US-based Highland Companies executives Joseph Izhakoff and John Scherer admitted that the company had always intended to quarry the land (Stiglic, 2011; Waters, 2011b).

When The Highland Companies began speaking publicly about their intentions for the quarry three dominant claims were presented. First, they downplayed the potential impacts of the proposed development on the watershed, agricultural capacity, and overall rural quality of life. For example, in the *Hydrogeologic and Hydrologic Assessment of the Pro-posed Melancthon Quarry*, the consultant predicted no significant impact on the surrounding groundwater and surface water as a result of the mega-quarry, a message later echoed by the proponent and its public relations firm (Genivar, 2011). In a statement that had been previously posted on the Melancthon Quarry website and is still referenced by NDACT, John Lowndes stated that:

> During the lifetime of quarry operation, and for as long as the water evacuation pumps require to be operational, the water table will not be significantly lowered, the adequacy of supply will not be comprom-ised and [The Highland Companies] will not contaminate the water supply making it unfit for consumption by all life forms.
>
> (Howard, 2011)

In its application for a licence, The Highland Companies' traffic scenario described 150 trucks driving to and from the quarry site per hour along Highway 124, totalling 7,200 passages a day (Bell et al., 2012). This volume of traffic along the haul route would undoubtedly impact other industries in Melancthon, such as tourism and agriculture but the proponent downplayed this repeatedly (Munro, 2011; Respondent 5, personal communication, 15 March 2016).

The Highland Companies used a 'reliance on science' framing (Zavestoski, Agnello, Mignano, & Darroch, 2004) as they assured the general public that reliable scientific tests and studies were being undertaken in accordance with Provincial policy and that a licence would not be issued by the Province unless MNR had confidence that their water management plan was sound (Waters, 2011b).

The Highland Companies emphasized in press releases and in the media that the quarry would create jobs in the community, estimating 465 permanent positions at peak production (Altus Group, 2010 as cited in Bell et al., 2012). In February 2012, The Highland Companies made a presentation to members of Council for Melancthon Township. They highlighted the fact that, should the quarry be approved, the Township would receive royalties for the aggregate extracted at a rate of six cents per tonne. This would amount to around $600,000 annually (Bell et al., 2012). The positive assessments were challenged in a variety of ways by the framing strategies and the actions crafted by the opposition.

Process

Shortly after the application for a licence was submitted in April 2011 mobilization against the quarry accelerated.

Oppositional framing

The social movement mobilized a diverse array of facts and values in their opposition to the quarry. Four general themes appeared: the threat to food and farming; the potential impact on the natural environment especially water; the lack of transparency on the part of the aggregate industry and government agencies; and the overall social cost that the project could have to life in the area near the quarry. All these themes echoed through the one and one-half years of intense opposition that began with the formal application for a licence. There were multiple entry points for the building of oppositional frames: water, soil, corporate power, quality of life, but food emerged as the central frame with a capacity to draw many thousands of Ontarians into the quarry opposition.

Food framing

In early 2011, NDACT members and other partners from Melancthon knew that it was crucial that they raise awareness about the proposed project and get it in the spotlight in order to highlight the lack of an environmental assessment in the regulatory approval process and to put pressure on the aggregate industry and the Provincial government. Movement leaders described how they realized early on in the mobilization that although they were indeed against the proposed development, it was better to be 'for' something rather than 'against' something. As such, NDACT and their supporters made a point of emphasizing the importance of local food and clean water. Slogans such as 'Taters not craters' and 'Save the land that feeds us' were effective at engaging urbanites in the City of Toronto.

> We had to take [the campaign] beyond the farm land and get it into the city, because Toronto chefs, and Toronto residents, eat a lot of the food that's grown in Dufferin County … So a bunch of us met, and it just so happened that a woman stepped forward who worked for the Food Network and she happened to know all of these top Toronto chefs. A bunch of us met in the city and she knew a restaurant that serves local foods, and so one of the first events that was organized was a mega quarry information session in downtown Toronto. The idea was … that the story had to be that 'this is the land that feeds us' … that mega quarry information session, which took place in June of 2011 at the Marben restaurant in downtown Toronto, attracted about one hundred people. We invited a whole bunch of people: there were bloggers, environmentalists, First Nations, chefs, the regular public influencers who had just come that night to drink some wine, have some local food, and hear the presentation. We had two farmers who came down from Melancthon, and we had a Toronto chef, and a Toronto food activist. And they just spoke. It was one of these old-fashioned salons with people just sitting and standing and listening to this incredible story.
>
> (Interview, 12 February 2016)

That information session, the first in a series of thematic information sessions aptly titled 'Downstream Downtown', was widely successful. In fact, it was at that event that the idea for other food-oriented events were formed.

Environmental framing

Due to the reach of the watershed, the outcome of the land use decision would impact the water resources of much of southwestern Ontario, and those involved in framing the environmental issues emphasized both the vast geography and the degree of risk (The Agenda, 2012; VanDyken, 2011). The David Suzuki Foundation, along with other environmental NGOs in Toronto, completed a lot of science and policy work to serve as a backstop to the environmental framing and give the movement credibility. It was thanks to this type of support and collaboration that the activists from Melancthon were able to avoid accusation of NIMBYism.

In submissions to MNR and MOE, independent technical experts expressed concern about the adequacy of the research presented by the Highland Companies (Frind and Associates, 2011; Natolochny, 2011). Frind and Associates argued that Genivar's report on water management failed to establish that the impacts on groundwater would not be significant. Frind, a hydrogeologist and Distinguished Professor Emeritus from the University of Waterloo is considered a pioneer in the field of quantitative groundwater science and groundwater modelling. Frind volunteered his time to review parts of the *Hydrogeologic and Hydrologic Assessment* for the proposed mega-quarry because of his interest in the region and his professional expertise. He found the model to be unreliable, with worst-case scenarios overlooked. He argued that the proposed rehabilitation scheme of dewatering in perpetuity was likely unrealistic and that alternate means of rehabilitation without water management were not included in the assessment. Finally, he asserted that impacts on the quality of both surface water and groundwater that feeds into the aquifers serving much of southwestern Ontario, was not adequately considered (Frind and Associates, 2011). This is just one example of local technical expertise volunteered in support of the opposition movement.

The Nottawasaga Valley Conservation Authority (NVCA) expressed its objection due to potential negative impacts on surface and groundwater, on a cold-water fishery, on terrestrial natural features, and anticipated erosion and flooding (Nottawasaga Valley Conservation Authority, 2011, as cited in Bell et al., 2012, p. 21). Other environmental claims included: the threat to the Bobolink bird and other endangered species; water quality; the threat to agriculture in the region, particularly the food producing farms adjacent to the proposed quarry; air quality with the increased truck traffic; and the feasibility of a very ambitious plan for restoring the land to production. A prominent demand made under the environmental frame was for a full environmental assessment of the quarry proposal.

Economic framing

Economic concerns were also expressed about the impacts on the local economy: the ambitious scale of the quarry proposal would threaten the viability of the Township's agriculture and tourism activities, which had an estimated impact of $100 million and generated 1,400 local jobs (Munro, 2011).

Many people in the tourism industry and the agriculture industry were concerned about the increased truck traffic that would occur if the proposed quarry was built. The passages per day that would be required to transport the aggregate to the market were a point of great concern because of the impact that this would have upon the state of built infrastructure, but also on rural quality of life, the tourism industry, and agricultural production (Munro, 2011).

NDACT argued that the jobs that would be created by the quarry would not be offered to area residents when their agricultural or tourism-based businesses suffered as a result of the truck traffic. Truck drivers would be hired from other communities who were not going to buy their groceries locally nor patronize local shops. The drivers would simply be passing in and out of the community but there would not be any spin-off industries as a result. So, the claim was made that these jobs would not be 'good' jobs that would help the local economy prosper. This counter claim was really effective at dampening the effectiveness of the job-creation narrative on the part of the proponent.

A Professor Emeritus at the Rotman School of Business at the University of Toronto conducted a business case study surrounding the proposed mega quarry, the results of which demonstrated the negative impact on existing industry in and around Melancthon Township should the quarry have been approved. Upon completion, he volunteered his time to drive out to Melancthon and present his findings to local officials (Respondent 1, personal communication, 12 February 2016). Municipal councillors expressed concerns that the new tax revenue generated by the proposed development would not be sufficient to cover the costs of maintenance and repairs to the regional roads (Chapman, 2012). NDACT and other partners made the claim that truck driving jobs are not local jobs and that this is a big problem for the local community (The Agenda, 2011).

Corporate misbehaviour claims

In addition to the food, environmental, and economic frames there was a continuous echoing of claims about the questionable intentions and bad moral character of The Highland Companies and the hedge fund Baupost. Media articles painted the proponents as the 'bad guys' in this underdog story or simply positioned them as rich, out of touch with rural realities, and unconcerned about destroying the environment. As one informant described the situation: There was a villain in Boston with all the money contrasted with a pristine and productive fertile plateau, an aquifer, and all of the food that it produces, water flowing to a million people downstream, and somebody wants to go and blow it up. For no reason other than greed. One Boston newspaper even did a story about how one of their citizens was involved in this land grab (Interview, 15 March 2016).

Consistent with this moral critique it was observed that although The Highlands Company was prohibited from working on the site after the Province announced in September 2011 that an environmental assessment would be required for the proposal NDACT reported that prohibited site alteration activities continued to take place (Bowman, 2012a). The company claimed in response that their activities were consistent with routine potato farming practices (Lowndes, 2012). Iler Campbell LLP, NDACT's legal representation, updated the Ministry of the Environment as to The Highland Companies' failure to cease site alterations as required by the environmental assessment process. NDACT stated their suspicion that the company would continue to push forward with the quarry application process (which had already garnered 5,000 objections) (Bowman, 2012b). NDACT reported that around March 2012, seven months after the requirement for an environmental impact assessment had been announced, company employees were clearing grasslands and brush from lands not suited for potato farming, clearing lands on road allowances, mulching areas known to be too wet for farming, cutting swales and ditches, altering the watercourse and demolishing heritage buildings (Bowman, 2012c). By describing the firm's contradictions and lack of transparency NDACT was effective at connecting many different stakeholders who held similar values surrounding the issue of industry transparency.

Framing social cost

Claims about the social cost of the project included: the challenges the proposal posed to community cohesiveness, as some community members were excited at the prospect of the quarry development while others were firmly against it; local governance issues; increased pressure on local agriculture; truck traffic that would decrease local quality of life; and food sovereignty.

Technical support for framing

The potential impact on urban southwestern Ontarian sparked important contributions from technical and scientific experts. Once allied with the movement's frames and values, they wanted to contribute what they could. Individuals and groups focused on water protection, on food security and food justice, and proponents of conventional agriculture were keen to contribute what they could in the effort to stop the quarry (Respondent 4, personal communication, 25 February 2016).

Technical expertise was important particularly to the environmental framing of the movement. Professionals, academics, and graduate students produced scientific, legal, and policy-related reports that challenged the proponents' technical studies. Technical experts who contributed their knowledge and time to the social movement included: law students and lawyers; engineers; hydrological modelling experts; land-use planners; GIS technicians; business professors; economists; conservationists; and soil health scientists. Interest groups allied with the opposition movement, such as the Ontario Federation of Agriculture and environmental NGOs, solicited a great deal of the technical research through their networks. Almost all of this expertise was volunteered; NDACT and other aligned organizations made a point of not paying for technical reports and studies if it could be volunteered. They found that experts were usually keen to support them by lending their time and knowledge to the cause. These served to give the movement legitimacy. Movement leaders all acknowledge that without this support, the outcome could have looked very different.

In addition to the David Suzuki Foundation's scientific contributions on the topic of farmland preservation, many academics hopped on board. NDACT engaged a soil specialist from the University of Guelph who completed soil studies for them free of charge. A GIS mapping engineer from a multidisciplinary engineering consulting firm was also an invaluable source of technical expertise. Some of Melancthon's residents could provide a wealth of professional knowledge as well; one resident was among the top land-use lawyers in the province and another was employed by MNR. These residents shared their expertise and it was critical to building capacity and legitimacy for the movement.

Many graduate students and professors helped with data collection and analysis. Graduate students of land use planning from the University of Guelph prepared a contextual report on the issue for the Ontario Federation of Agriculture, and a group of law students from Osgoode Hall Law School came together to support NDACT with pro bono legal expertise (Bell, et al., 2012).

Using social media

As one interviewee observed social media was used effectively. Movement participants attended rural events and compiled emails and lists using these to keep the constituency informed about activities and people through regular updates. The constituency grew with an estimate from one organizer of close to half a million people following on social media. The focus was on the Greater Toronto Area. Alliance building was supported by celebrity endorsements. Rachel McAdams a well-known Canadian actress endorsed NDACT's Food & Water First campaign and Margaret Atwood an internationally acclaimed writer retweeted NDACT tweets. This helped to spread the message that the opposition movement could be trusted and that Baupost and The Highland Companies could not be.

NDACT, was one of the primary organizations involved in the movement. When taking human resource capacity, expertise, and influence into account, it did not try to overly control all activist actions and messaging. NDACT worked to keep the movement upbeat and positive, despite the anger directed at the proponent, inter-organizational disagreements, and growing pains. This is evidenced in how their mandate evolved over time. While individuals initially began meeting with one another out of a common concern over what the proponent might be planning for the land, NDACT quickly shifted its core message from one of opposition to aggregate development to being *for* food and water (Van Bruggen, 2012).

Actions

In addition to the oppositional framing the movement generated some important and innovative actions including the First Nations Walk, Downstream Downtown, Foodstock, and Soupstock.

First Nations Walk

An early event was the First Nations Walk which was organized over the Easter weekend of 2011. A group of farmers, ranchers, and First Nations leaders organized to walk the 120 kilometres from Queens Park in downtown Toronto, the seat of the Provincial government, to the quarry site. Nearly 200 people gathered at Queen's Park for the send-off, where speakers highlighted the potential risks to water supply, agriculture, and rural quality of life. Over the course of the next five days, over 1,000 people were involved in the march and the story was picked up by most major national television stations, local radio, and newspapers. This event set the tone for the inclusive and collaborative movement that began to grow and the event crystalized a few of the key claims that were emphasized to the media, government representatives, and the general public throughout the months that followed (Waters, 2011a). A prominent demand was for an environmental assessment of the proposed quarry.

Downstream Downtown

On 28 June 2011, Downstream Downtown was held at the Marben Restaurant in Toronto. This meeting of farmers and landowners with food activists and citizens attracted politicians, journalists and broadcasters, as well as some of Toronto's top chefs. The event dealt with food security issues associated with the Melancthon development. A second Downstream Downtown event focusing on water was held in September 2011 at Patagonia, a Toronto store that specializes in outdoor apparel. Maude Barlow of the Council of Canadians, an internationally known water activist, spoke about the proposed quarry. These drew urbanites, who were supportive of local food and environmentally-focused, as well as policy experts, and technical experts.

Between April and September 2011, the Minister of the Environment received more than 700 letters requesting an environmental assessment (Ministry of the Environment, 2011, as cited in Bell et al., 2012, p. 9; CORE, 2011; Keller 2011). Responding to this pressure on 1 September 2011, it was announced that the proposed quarry would be subject to a full environmental assessment in accordance with the Environmental Assessment Act (Ministry of the Environment, 2011, as cited in Bell et al., 2012, p. 9). As of that date The Highland Companies was prohibited from undertaking any site alterations related to the proposed project without the approval of the Minister. The requirement that an environmental assessment be conducted was an important early victory but was only a step in the mobilization against the mega-quarry.

Foodstock

Within a month of the announcement that an EIA would be required NDACT, the Canadian Chefs' Congress, and a number of other partners in the quarry opposition hosted a culinary protest entitled 'Foodstock'. The event gathered local chefs from across the country, famous musicians, and over 28,000 people on the fields surrounding the proposed quarry site (Elton 2011; Stadtländer, 2011). The event was pay-what-you-can and organizers emphasized their desire for the event to be inclusive, fun, celebrating clean water and a bountiful harvest (Bain, 2011; Stadtländer, 2011). Intermittent rain showers on 16 October did not appear to dampen the spirits of the culinary pilgrims, outfitted with boots and armed with their own cups. Thousands made the hour's drive from Toronto and many more arrived on chartered buses arranged from Toronto, Guelph, Hamilton, and Collingwood. At one point in the day, traffic sat bumper-to-bumper from the farm gate to the Town of Shelburne, 13 kilometres away (Waters, 2011c). Foodstock through concerts and culinary attractions brought Torontonians out to learn about the quarry issue.

With Foodstock, 28,000 people flooded into a community of 2,800. People from Toronto came all that way so they could eat food in the forest.

> They were cold and wet but they were under a huge sky, and they felt they were involved in something different. There's something authentic to be out there on the land. It was absolutely amazing and that really galvanized people.
>
> (Personal communication, 12 February 2016)

What Foodstock did for the local community, and even more so for those who were marginally involved or marginally interested in the movement, was to get them out on the land, eating the harvest grown there, and listening to the stories of those who would be impacted by the proposed development. This raised awareness and support for quarry opposition efforts, raised funds for legal costs, and celebrated the recent victory of the mandated environmental assessment.

Soupstock

One year later on 21 October 2012, a second major culinary and information event, Soupstock, was held in downtown Toronto, at Woodbine Park. It was organized by the Canadian Chefs Congress and the David Suzuki Foundation and featured soup prepared by 200 prominent chefs, 85 different soup stations, guest speakers and musical entertainment. Over 11,000 lbs. of food were donated. It attracted an estimated 40,000 participants (Loney, 2012). There were some people who explicitly came out to learn about the issue but many more were drawn out by the fun activities; listening to some great Canadian bands and eating great food. It was all free so people had nothing to lose. This event also generated donations to support the work of NDACT.

Outcomes

On 21 November 2012, almost three years after the initial meeting at the Honeywood arena and one and a half years after the quarry application had been submitted to MNR, the application for the extraction licence as well as the plans for a rail corridor to Owen Sound were withdrawn by The Highland Companies. A spokesperson told the media that the company realized that 'the application does not have sufficient support from the community and government to justify proceeding with the approval process' (CBC News, 2012).

Preservation of the farmland was the first important outcome of the movement. On 16 July 2013, Bonnefield Financial, an investment firm specializing in land acquisition to preserve prime foodland, purchased all of the company's land. Bonnefield's president, Tom Eisenhauser, announced the fields will continue to be used for food production (Overview on NADCT, 2013).

A second outcome has been the institutionalization of NDACT which continues to engage its Food & Water First campaign mobilizing for farmland preservation in Ontario. Food & Water First has continued to engage southwestern Ontarians, especially urbanites, on local food economies and supply chains. There is now a greater awareness about the importance to urban areas of rural exports and in fact, the mutual interdependence of urban and rural areas.

The Food & Water First Movement has continued to work to reform land use policies. Food & Water First as an organization has been actively involved in the ARA Review process and continues their presence in the media and through social networking. The ongoing engagement with politicians and policy makers through the Food & Water First campaign has put a spotlight on the way that politicians are voting on food and aggregate issues and on the reliability of those politicians in following through. Transparency is promoted through their strong and continued social media presence. Food & Water First has continued to broadcast these issues, tweet about them and get this information out to people. They've helped to provide greater transparency around what's actually happening in government offices and at Queen's Park (Respondent 3, personal correspondence, 24 February 2016; Respondent 5, personal communication, 15 March 2016).

Another outcome of the movement has been an increase in the integration of the community in Melancthon Township. Before the opposition movement the community was quite divided between those whose families had lived and farmed for generations, and the weekenders who were migrating to the community from urban areas: musicians, educators, lawyers, and others with an urban base. There was little intermingling between these two distinct groups. However, the opposition movement knit the long-time residents and urban weekenders together permanently transforming the community's identity. In the words of one Melancthon resident, 'if the movement did anything, it was to create a stronger community' (Respondent 5, personal communication, 15 March 2016).

The activism connected the rural community with policy-makers, press, and activists in Toronto. We see to this day a lot of communication and collaboration between the weekenders and the farmers, as well as all those most closely involved in the opposition effort; they have stayed engaged with one another though the continued work of NDACT under the Food & Water First banner. They still meet throughout the year and several interviewees made a point of explaining that there are such rich memories and a high degree of loyalty that this network can now draw upon.

Conclusion

The successful mobilization against the Melancthon Mega-Quarry is a rare example of success in blocking a large-scale resource development project promoted by a corporation with substantial capital to marshal support against community opponents. The success was a product of several framings which may be quite unique to this single case but that might also provide strategic lessons for other movements. The focus on food proved to be the central framing success. Key events included meetings at restaurants, on farms and in parks. Downstream Downtown, Foodstock, and Soupstock pulled in that segment of the general public who shared an interest in food as a cultural commodity and as a contributor to health. The close proximity to Toronto with its active and high-profile food scene and its very active local food movement attracted urban support for the quarry opposition. Toronto is a large urban centre with a population of 2.5 million. By framing the quarry as a development in Toronto's backyard and a development that threatened Toronto's food, the urban population was drawn into what might otherwise be viewed as a purely rural and local development issue. Recruitment of chefs, musicians and artists with a national profile reinforced the urban interest. The food theme was echoed by linking food and farming, food and water, food to the environment. The environmental framing supplemented and intensified the food framing and was the most important site for the contribution of technical expertise. The range of environmental analysis and documentation that was provided gave credibility to the claims over the threat to water. Other frames stressing corporate greed and dishonesty or the potential impact on local social values were perhaps less important to the widespread opposition and the eventual decision of The Highlands Company to abandon the project. What began as a local development proposal grew into a provincial and national opposition through months of dedicated movement work. At the core of the success was a deep cultural identification with food and farming that reached far beyond the local area on which the proposed quarry would be developed. The Stop the Quarry movement generated a widespread provincial and national opposition that achieved its central aim and continues to mobilize for farmland and water protection in Ontario and across Canada.

Note

1 Upon withdrawing their application for licence, The Highland Companies and their public relations firm, Hill & Knowlton, erased their digital footprint. All webpages disappeared and there is a no lingering record of their social media presence, although newspaper articles, correspondence, and press releases are still available from other sources. As such, there are limitations on the scope of the analysis as many of the claims utilized to shape the discourse are limited or missing. The documents collected included: newspaper articles from print and online newspapers; selected blog posts written by key movement leaders; legal correspondence from representatives of both The Highland Companies and the opposition movement; technical and scientific reports; technical reports prepared by academics and graduate students; press releases; government documents; and video clips of interviews for television that have been posted on YouTube and Vimeo. The breadth of media coverage and grassroots research that occurred between 2009 and 2012, as well as the disappearance of much of The Highland Companies' material following the withdrawal of their application, precludes the possibility of collecting all relevant data sources. Every effort has been made by the authors to ensure that a wide cross-section of the published material was collected and used as part of this study. The Highland Companies published their official statements on two websites: www.melancthonquarry.ca and www.highlandcompanies.ca. Both websites have since been taken down and their contents are irretrievable. The latter domain name has since been taken over by a private individual who created a spoof site.

References

Agenda, The. (2011, 28 October). *Mega Quarry, Mega Worry*. The Agenda with Steve Paikin. Video file. Retrieved 22 January 2016 from www.youtube.com/watch?v=ZTTR-Fer0aE.

Agenda, The. (2012, 26 November). *How the Mega Quarry Fight Was Won*. The Agenda with Steve Paikin. Video file. Retrieved 22 January 2016 from www.youtube.com/watch?v=rT-auyCpBo4.

Aggregate resources (n.d.). Province of Ontario. Retrieved 22 January 2016 from www.ontario.ca/page/aggregate-resources#section-0.

Altus Group. (2010). *Economic Benefits of the proposed Melancthon quarry and financial impact on the township of Melancthon*. Originally posted on www.highlandcompanies.ca/index.php/companies/melancthonquarry/technical_studies.

Bain, J. (2011, 17 September). Chefs, farmers unite for Foodstock. *The Toronto Star: Food and Wine*. Retrieved 16 February 2016 from www.thestar.com/life/food_wine/2011/09/17/chefs_farmers_unite_ for_foodstock.html.

Bell, D., Isaac, J., Jamal, A., Levay, D., and Wright, A. (2012). *Assessing the impact of the Melancthon quarry: Prepared for the Ontario Federation of Agriculture*. Guelph, ON: School of Environmental Design and Rural Development, University of Guelph.

Binstock, M. & Carter-Whitney, M. (2011). Aggregate Extraction in Ontario: A Strategy for the Future. *Canadian Institute for Environmental Law and Policy*, pp. 1–78. Retrieved 22 July 2016 from http://cielap.org/pdf/AggregatesStrategyOntario.pdf.

Bowman, L. (2012a). 'Highland Companies Site Alteration and Related Activities'. Letter to The Highland Companies. 1 March 2012. Unpublished manuscript.

Bowman, L. (2012b). 'Highland Companies ARA Application'. Letter to the Honourable Jim Bradley, Minister of the Environment. 21 March 2012. Unpublished manuscript.

Bowman, L. (2012c). 'Highland Companies Site Alteration and Related Activities'. Letter to John Lowndes. 21 March 2012. Unpublished manuscript.

CBC News. (2012, 21 November). 'Mega-quarry' in southern Ontario won't be built. Retrieved 15 April 2016 from www.cbc.ca/news/canada/toronto/mega-quarry-in-southern-ontario-won-t-be-built-1.1187522.

Chapman, S. (2012, 10 July). Farmers revolt. *Toronto Life*. Retrieved 15 April 2016 from www.ndact.com/index.php/media-stories/media-stories-magazines/498-farmers-revolt-toronto-life-magazine.

CORE – Conserve Our Rural Environment. (2011). CORE objection letter to MNR. Unpublished manuscript.

Devlin, J. F. (2006). Why is environmental assessment failing? Political contention, public participation and the limits of democratization. Presented at the IPSA-AISP Conference in Fukuoka 2006. Retrieved 25 June 2016 from http://paperroom.ipsa.org/papers/paper_5304.pdf.

Devlin, J. F., Yap, N. T., & Weir, R. (2005). Public Participation in Environmental Assessment: Case Studies on EA Legislation and Practice. *Canadian Journal of Development Studies*, 26, 3, 487–500.

Devlin, J. F. & Yap, N. T. (2008). Contentious politics in environmental assessment: blocked projects and winning coalitions. *Impact Assessment and Project Appraisal*, 26, 1, 17–27.

Elton, S. (2011). Foodstock: Canadian Foodies and Chefs Fight Mega Quarry in Ontario, *The Atlantic*, 27 October. Retrieved 25 March 2016 from www.theatlantic.com/health/archive/2011/10/foodstock-canadian-foodies-and-chefs-fight-mega-quarry-in-ontario/247339/.

Frind and Associates Ltd. (2011). 8 July 2011. Letter to Ministry of Natural Resources and The Highland Companies. Retrieved 3 July 2016 from http://66.212.167.146/MelancthonMegaQuarry/pdfs%5CE_ Frind_Letter_Melancthon_ Quarry_July8-2011.pdf.

Gayler, H. J. (1982). Conservation and Development in Urban Growth: The Preservation of Agricultural Land in the Rural-Urban Fringe of Ontario. *The Town Planning Review*, 53, 3, 321–341.

Genivar Inc. (2011). *The Highland Companies Proposed Melancthon Quarry Hydrogeologic and Hydrologic Assessment, Volume 1 of 4, Summary Report.* Montreal: Genivar.

Howard, A. H. (2011, 11 September). Response to John Lowndes, 'An update on the proposed Melancthon quarry'. NDACT. Retrieved from 20 April 2016 from www.ndact.com/index.php/component/content/article/33-letters-reports-sitings/letters-of-objection/132-response-to-john-lowndes-qan-update-on-the-proposed-melancthon-quarryq.

Keller, W. (2011). Highland Companies wants rail line. Blog Post. Retrieved 14 February 2016 from http://orangevillemarketwatch.typepad.com/thewatersgroupjournal/2011/05/highland-companies-wants-this-rail-line-13-hours-of-trains-every-day-in-orangeville.html.

Loney, H. (2012). Tens of thousands attend Toronto's Soupstock, mega-quarry protest. *Global News* 21 October 2012. Retrieved 21 March 2016 from https://globalnews.ca/news/299452/tens-of-thousands-attend-torontos-soupstock-mega-quarry-protest/.

Lowndes, J. R. (2012). Response to 1 March 2012 Letter. Letter to Iler Campbell LLP. 9 Mar. 2012. Unpublished manuscript.

Ministry of Finance. (n.d.). *Ontario Population Projections* Retrieved 20 March 2016 from www.fin.gov.on.ca/en/economy/demographics/projections/.

Ministry of Natural Resources. (2010). *State of the Aggregate Resource in Ontario Study.* Toronto: Ministry of Natural Resources.

Munro, R. (2011, 30 June). Hills of the Headwaters Tourism Association letter to the Minister of the Environment. Retrieved 3 July 2016 from http://66.212.167.146/MelancthonMegaQuarry/pdfs/HillsHeadwater_30-June-2011.pdf.

Natolochny, F. (2011). 'The Highland Companies Proposed Melancthon Quarry Application'. Letter to Craig Laing, Aggregate Resources Officer. 26 April 2011. Unpublished manuscript.

NDACT History (n.d.). Retrieved 20 March 2016 from www.ndact.com/index.php/about-us/history.

NDACT's Mission Statement (n.d.). Retrieved 20 March 2016 from www.ndact.com/index.php/about-us/mission-statement.

Overview on NDACT (2013). Retrieved 20 March 2016 from www.ndact.com.

Shuff, T. (2011, 16 June). Melancthon mega quarry by the numbers. *In The Hills*. Retrieved 22 March 2016 from: www.inthehills.ca/2011/06/back/melancthon-mega-quarry-by-the-numbers/.

Stadtländer, M. (2011). *Foodstock sponsorship package*. Retrieved from 18 March 2016 from http://orangevillemarketwatch.typepad.com/Foodstock/Foodstock%20 Sponsorship%20Package.pdf.

Stiglic, J. (2011, 10 October). Quarry on Ont. farmland was the plan, firm says. *CBC News*. Retrieved 20 April 2016 from www.cbc.ca/news/canada/quarry-on-ont-farmland-was-the-plan-firm-says-1.1077670.

TransAlta (2016, 12 January). *Melancthon*. Retrieved 2 April 2016 from www.transalta.com/facilities/plants-operation/melancthon.

Van Bruggen, J. (2012, 23 November). How the war against the mega quarry was won. *The National Post*, Retrieved 15 March 2016 from http://news.nationalpost.com/full-comment/jason-van-bruggen-how-the-war-against-the-mega-quarry-was-won.

VanDyken, R. (2011, 9 December). Melancthon mega-quarry. Water concerns and environmental risk assessments impede plans for a mega-quarry near the GTA, *Canadian Geographic*. Online. Retrieved 18 April 2016 from www.canadiangeographic.ca/article/melancthon-mega-quarry.

Waters, D. (2011a, 22 April). Activist and Area People Embark on 5 Day Walk to Protest Mega Quarry [Web blog post]. Retrieved 17 April 2016 from http://orangevillemarketwatch.typepad.com/thewatersgroupjournal/2011/04/activists-and-local-area-people-embark-on-a-five-day-walk-to-protest-mega-quarry-its-all-about-quail.html.

Waters, D. (2011b, 10 October). *CBC's The National Covers the Mega Quarry and Foodstock*. [LivingInOrangeville Video file]. Retrieved 17 April 2016 from www.youtube.com/watch?v=blI6sMKnHoE.

Waters, D. (2011c, 21 October). Foodstock 2011 updates and memories [Web log post]. Retrieved 17 April 2016 from http://orangevillemarketwatch.typepad.com/thewatersgroupjournal/huge-quarry/.

Zavestoski, S., Agnello, K., Mignano, F., & Darroch, F. (2004). Issue framing and citizen apathy toward local environmental contamination. *Sociological Forum*, 19, 2, 255–283.

9 Rural protests and the mining industry in Finland

Tuija Mononen and Ismo Björn

Introduction

Mining is typically ruled by large-scale multinational enterprises (Moody, 2007; Saleem, 2009). In the global economy, mining activities can be understood as a complex struggle between global and local actors for the utilization of local natural resources (Massey, 2008; Singh & Evans, 2009). Differing views and divergent values may lead to conflicts and stimulate activism and protests, and to the emergence of anti-mining movements. Often, the main concern is connected to the quality of everyday life and the environment, and environmental impacts have caused concerns in many mining locations. Impacts of mining on water systems cause concern generally and the risks associated with wastewater management are a source of conflict between local residents and the mining companies. Water availability, potential water pollution and the safe use of water may all raise concern among local residents (Conde & Le Billon, 2017; McKenzie, 2009). Local impacts of mining differ between mining projects, and countries. Impacts can be environmental, economic, political, and cultural, and can affect the everyday life and wellbeing of local people. The impacts can be perceived as bringing wealth to regional economies and local communities, but they also may be perceived to harm the environment, the inhabitants, and their livelihoods. In many locations, particularly in remote rural areas, mining may be the only promising option for development (Measham, McKenzie, Moffat, & Franks, 2013; Mononen, 2012, 2015; Mononen & Suopajärvi, 2016; Richards, 2009).

In the past, Finnish mining was primarily in the hands of the state. Outokumpu Ltd. was one of the major state-owned operators. With strengthening global demand for minerals and with the growth of the mining industry there was an upswing of mining in Finland in the first half of the 2010s. Today, the actors in the field are mainly foreign corporations expanding their activities increasingly in remote areas. With this increase, discussion and activity concerning the use of natural resources increased. Finnish rural and regional policy has considered the mining industry highly promising because it is supposed to have positive socio-economic impacts on rural areas both directly and indirectly (Saartenoja, Törmä, Valkosalo, & Zawalinska, 2007). In Finland, concerns about mining have been based on the rapid expansion of mining, the entry of foreign mining companies, and failures in the management of environmental impacts. There are movements opposing mining in different parts of Finland and they are interrelated, learning modes of action from each other and from older traditions of rural protest (Björn, 2016a, 2016b).

This chapter is based on several qualitative research projects connected to three mining cases in Finland which have given rise to opposition (Figure 9.1). All three mines are located in remote rural areas in Eastern Finland. In all cases, there have been local worries about environmental impacts especially those connecting to the management of wastewaters. The first, Outokumpu, was remarkable economically and socially, and had

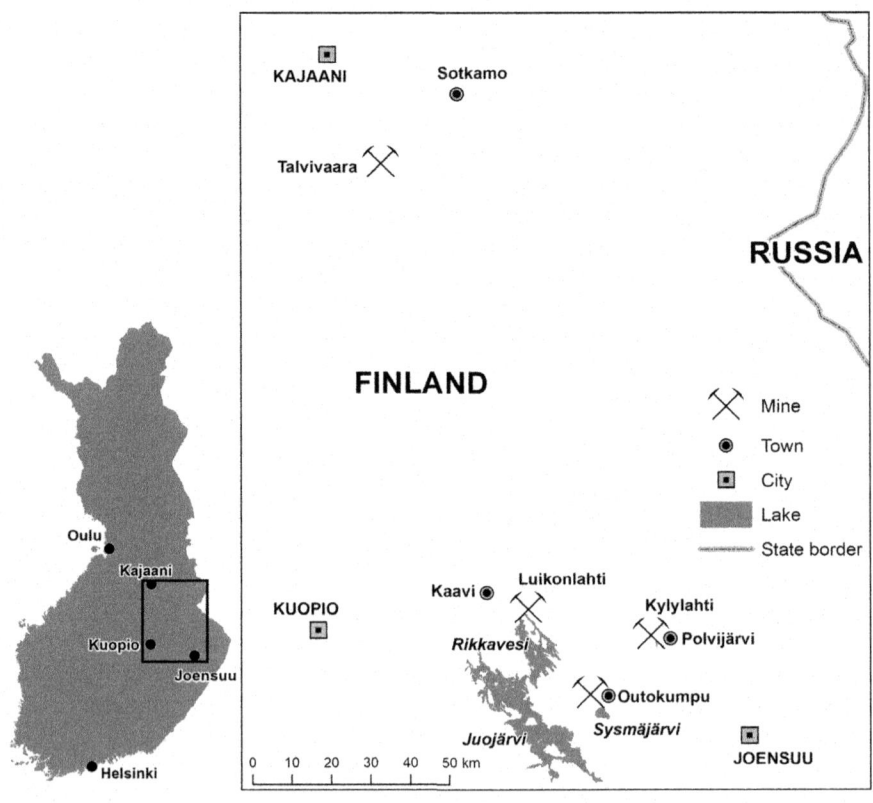

Figure 9.1 Locations of Outokumpu, Talvivaara, and Kylylahti mines.

Source: Map by author Ismo Björn.

many environmental impacts that still affect attitudes towards mining in Finland today. In the second case, Talvivaara, environmental impacts led to conflict that has evolved from a local environmental conflict to a national conflict symbolic of the mining industry throughout Finland (Tiainen, Sairinen, & Mononen, 2014; Sairinen, Tiainen, & Mononen, 2017). The third case, Kylylahti, is similar but more recent with environmental concerns leading to opposition.

Interviews with various stakeholders, including local residents and entrepreneurs, representatives of mining companies, municipalities, inspection and environmental organizations, combined with the results of the analysis of other data (newspaper articles, historical documents) have been used to provide an analysis of local activism connected to the mining industry. Other related literature has been used as background data and in case analysis. We explore how activism and protest groups are mobilized, how they are organized, and who are the key actors. We describe what repertoires of action have been used. We also consider how mining companies and other relevant stakeholders have reacted to the resistance. Qualitative content analysis has been used. The main questions and data have been analysed drawing primarily on the Finnish literature.

Theoretical frame

The increase in civic protest and activism in rural areas is an international phenomenon. In the international rural research literature protests and analysis of social movements has become more intense (Scott, 1985; Strijker, Voerman, & Terluin, 2015; Woods, 2008, 2011). Research on anti-mining movements is expanding as well (Urkidi & Walter, 2011). By protesting rural people are defending or opposing for example construction of a new road, a wind turbine park or new residential areas. The protest itself can be connected to one issue, but usually it is related to the broader defence of the rights and autonomy of rural residents (Björn, 2003). Local residents may experience their traditional rural way of life threatened by some external or global phenomenon. More broadly, protests can be interpreted as a defence of the meaning of the rural and the use and rights of rural areas (Woods, 2003).

The protests may be local, but the same phenomenon may appear in more than one place depending on the economic, cultural and political history of the area (Larsen, 2008; Woods, 2008). Protests differ in scope, duration, location, as well as political impact. The ideological background and strategies vary, and the spread of protests may follow a pattern with peaks and troughs. Many of the protests experienced in Finnish rural areas have been framed as environmental issues. In many cases there have also been outside actors involved (Björn, 2003; Rannikko, 2003). Mining movements are taking advantage of a versatile

tradition of collective action. Collective action does not always constitute a social movement. The duration of action varies, and short-term
eruptions may not be sufficiently sustained to be considered a social
movement. Social movement – as an analytical concept – refers to forms
of collective action with common objectives and temporal continuity.
Social movements have an interactive network and their own identity,
and they are often active outside the channels of political representation. They are changing and dynamic and their temporal continuity does
not require a formal organization (Rasimus, 2006). The concept of the
social movement is quite precise. Still, the use of the concept has been
patchy and it stretches to a wide range of descriptions of organizations
and contexts.

Protest models are international and common ways to protest include
seminars, meetings, petitions, complaints, demonstrations, and civil diso-
bedience (Konttinen & Peltokoski, 2004). Although every movement is a
product of its own political culture and of the conditions of a certain
period, the principles of resistance and the processes by which a movement's political base can be constructed remain largely the same (Scott
1985). Cyclical vicissitudes are inherent to societal activism (Rannikko,
2003). They tell about the conflicts and dissent characteristic of a specific period.

In the past, the most conspicuous representatives of agricultural organizations and defenders of rural areas were farmers. Rural interests were at
that time equal to the interests of agriculture and farming organizations,
and the primary goal was to defend the economic interests of farmers
(Woods, 2003). The agrarian movement can be classified as an 'old'
social movement. Its target included the quest to protect the social interests of the one societal group (Konttinen & Peltokoski, 2004; Rasimus,
2006). The birth of new social movements is often associated with the
rise of soft values in society. The motives of the actors are based for
example on questions of daily life and the quality of life. The focus has
moved away from material interests and new movements have been
described as socio-cultural or identity movements (Rasimus, 2006).
Woods (2003, 2008) has also noted that many of the more recent groups
in rural areas focus on interests connected to rural identity and quality of
life. Newer groups do not necessarily want to be involved in official
decision-making, while the older groups are usually part of local political
decision-making bodies. Newer groups are interested in action that gets
public attention and they are often interested in defending rural areas
from an external threat. They are often informal in nature, emotional,
short-term, and operate at a grass-roots level. There is not necessarily an
obvious group leader (Rasimus, 2006). In Finland, civic activism has
traditionally based on formal organizations (Lindholm, 2005; Rasimus,
2006) but with mining civil society actors have emerged outside the
system of formal representation.

Activism around the mining industry – three Finnish cases

Outokumpu and the wastewater dispute

Outokumpu mine was located in North Karelia in Eastern Finland (Figure 9.1) and operated from 1910 to 1989. The mine gave its name to the current location. The ore was rich containing copper, iron, sulphur, zinc and cobalt, and a small quantity of gold and silver. The mine employed up to 1,100 employees (Kuisma, 1989; Särkikoski, 1999). Because of a short-lived copper mill and the early ore beneficiation process (1913–1929) the area around Outokumpu mine was heartbreakingly black. Waste and ground water problems affected a large area outside the mine. The question of ground water shook the whole community, but especially the rural population. The company's management, interaction, and communication were severely tested (Björn, 2000; 2016b).

Ground water and surface water near the mine were polluted by waste-water. The Outolampi pond was a dump for the wastes of the processing plant. First, tens of thousands of tons of waste liquid was allowed to run into the pond. The mining company's brochure 'Outokummun kaivos 1910–1928' suggested that the pond would be like 'a container, where the waste can sink to the bottom, so that only the clear water will run into another water body' (Björn, 2014, p. 138). In this wastewater there was sulphuric acid and cyanide-containing flotation chemicals in addition to solid waste. The water pumped from the old mine contained heavy metals and acidic water. Various types of wastewater and reactions between them could not be controlled even theoretically. Dams did not stop the waste-water which spilled through the dams. Part of the wastewater flowed into the soil.

In 1918, local people expressed for the first time their concern about the environmental degradation. Over the years, concerns grew and local people asked to talk with the director of the mine, but he refused. The company's approach to the environmental questions and to local people was considered rude (Björn, 2000, 2014). Local people as well as the agricultural and forestry sectors in the area were suffering from the emissions. The well water was undrinkable. Outokumpu Ltd. was planning to empty a whole lake in order to remove the waste liquid that was accumulated in the bottom of the lake. There were also experiences with overflows of dam ponds in the 1940s. Swimming was banned in a nearby lake in the 1940s because people swimming in the lake developed a rash. The water was acidic and one lake died rapidly (Björn, 2016a, 2016b). Only in December 1948 did authorities at last issue a statement that groundwater near the mine was unusable. A few years later Outokumpu Ltd. was ordered to arrange clean water for local people and as a result, the company built kilometres of water pipes and arranged for an alternative water supply. The bitterness did not calm down. The company's liability was clear but to get compensation local people were forced to litigate with the company (Björn, 2014).

The situation came to a head when cyanide was found in the ground-water near Outokumpu in the early 1960s (Ryhänen, 1960). It was taken for granted that cyanide had travelled from the wastewater and entered the ground water. But the authorities did not seem to consider the activities of the mining company to be a problem. The mining company, in turn, tried to prove, that the citizens who presented the charges connected to cyanide were themselves disturbed. The company announced publicly the names of local persons who made complaints. The mining company sought to show that those raising the complaints opposed the progressive mining industry. According to the company, the groundwater analysis was made by an untrustworthy method. In addition, they argued that the impacts of waste-water could not even reach to the area from which the original sample was taken. Authorities assured the community that the amounts of cyanide did not give rise to concern. Figures and numbers were presented to local people to indicate that there was no danger. When the water rights court sentenced the company because of the cyanide the company appealed to the Supreme Administrative Court and was found not to be responsible. Outokumpu Ltd. was only ordered to monitor the concentrations of cyanide in the groundwater (Björn, 2016a).

According to the company's information, the amount of cyanide in the soil was normal or it was not possible to demonstrate reliably the source of the increase in the cyanide levels. Technical information was offered to calm down the criticism (Huttunen & Sivonen, 1974). Local people made numerous complaints about the wastewater issue in the 1950s. The residents continued asking for their rights. The company's experts tried to minimize the inconvenience local people experienced and appealed to technical information. According to the director of the Institute of Agriculture, farmers themselves were to blame because they left ditching untreated and fertilized their fields too little. It was suggested that one source of cyanide was a high water table (Björn, 2016a). The National Board of Health was also asked to investigate the wastewater issue. However, it did not make an investigation instead it appealed the research made by the Agricultural Board. The Board's opinion was that there is no reason for fear.

In the region of Eastern Finland, a water law review was held in 1964, but the landowners were not satisfied with the definition of the mission. Confidence in the Board of Water Law faltered. In the opinion of land-owners, the statements given by the Board of Water Law seemed to support the company's line. The owners were demanding from the authorities and Outokumpu Ltd. protection of their rights (Björn, 2016a). The Deputy Chancellor of Justice recommended that the company treat the wastewater to take care of the issue. Outokumpu Ltd. introduced chemicals intended to neutralize the wastewater. It was an experiment. The company had tested various methods to decontaminate wastewaters in laboratory conditions and sought to learn from foreign mines. Neutralization of ferrous waste-water with lime was proven to be the least expensive to implement. Ten to

14 tons of lime were added to the wastewater each day. In this way, it was believed that Lake Sysmäjärvi could be saved. However, the Management of Outokumpu mine was charged with polluting waters on purpose or at least in a grossly negligent way. But in a pamphlet published by the company, three Professors of law at the University of Helsinki gave opinions suggesting that the top management of Outokumpu Ltd. could not be prosecuted because they were not responsible for the water pollution (*Vesistön pilaantumista* ... , 1966).

In the media, requirements of compensation made by nearly 300 landowners were described as a campaign against the company. The company argued that the mine's wastewater effluents were very different than the wastewaters of paper factories and municipalities because the mine's wastes did not create oxygen demand. According to the company, nothing illegal had happened because it was only in 1962 that the water law clearly forbade pollution of groundwater. When the company finished putting wastes into the old waste-dumping area in 1954 the pollution of groundwater ended and, according to the company, pollution no longer occurred. In this way, the company used the law to avoid its responsibility (*Vesistön pilaantumisesta* ... , 1966). The mining company was able to argue that they had acted legally. The mining company appealed to its own technical studies and to the amounts of money required for reimbursement, but also to the important role of the mining industry and its employment potential impact in Finland (Björn, 2016a). Outokumpu Ltd. played two cards. They offered compensations to landowners and local people to show good will while trying to prove the company's innocence in court at the same time. In the fall 1966, after a long court process, it was agreed the company would pay financial compensation to landowners and the company would also compensate for the damages to the environment (Björn, 2016a).

Talvivaara mine and the floods of mining waters

Discussion of the mining industry in Finland since 2010 has concentrated mainly on one case, Talvivaara multi-metal mine, now known as Terrafame. In 1977, the Geological Survey of Finland started research in the Talvivaara area, located in Kainuu (Figure 9.1). A mining concession was granted to Outokumpu Ltd. in 1986, but the company did not believe that exploitation of the Talvivaara deposit by conventional metal enrichment technologies was commercially viable. In 2004 Outokumpu Ltd. sold the mining concession for one euro to the new Talvivaara Mining Company Plc. There were great expectations from the perspective of employment, regional development and economy, and the project was described very positively in news headlines.

As part of the sale, the new company acquired testing and research information about the bio heap leaching technique. Metals are separated from the ore by means of enriched microbes. This method enables economic nickel extraction from low-grade ore. The process is slow, but it can be accelerated by using high amounts of sulphuric acid. The mining company reported that the method is an 'environmentally friendly process and based on natural processes' (Mononen, 2015; Tiainen, et al., 2014). Despite this assurance nearby residents began to express their concern about the accuracy of the assessment of the environmental impact of the mine in 2005 prior to the granting of an environmental authorization and a water permit and the initiation of mining operations. At that time the volume and impacts of mining on the river basin were considered to be unpredictable (Tiainen et al., 2014). Local residents were concerned about the environmental impacts, especially estimated impacts to lakes near the mine.

In June 2009, the mine was officially opened although already in 2008, only 18 months after construction had started, Talvivaara was able to deliver its first metal products. Predicted lifespan of the mine was over 60 years. When the mining started, odors from hydrogen sulphide emissions were detected within a radius of several kilometres. Odor nuisance was in the news: 'Talvivaara mine stinks despite promises'. In February 2010, Talvivaara announced new plans to recover uranium as a by-product. This aroused intense debate as to whether the company management had actually planned uranium production from the beginning. Soon after the announcement there was the first leak from the mine's gypsum sediment pond although the leakage did not cause emissions outside the mine's area (Tiainen et al., 2014).

During the fall of 2010, the mining company discovered that the sodium, sulphate and manganese content of its wastewater had risen considerably above the regulated limits. The main cause was the change from water to lye gas scrubbers which had been introduced to control odor. The solution to the odor problem had resulted in salinization of the local lakes. The mine's wastewater contained so much sodium that local lakes had turned salty (Tiainen et al., 2014; Mononen, 2015). There were discussions of symptoms of skin disorders. The CEO of the mine denied this by saying that there were no proofs about the pollution of the environment. The mine was the subject of extensive criticism and the number of actors engaged in the debate expanded. In September 2011, the regional environmental authority (The Kainuu Center for Economic Development, Transport and Environment, KaiELY) made a request to the police to investigate the treatment of mine effluents. The local fishing association and the water cooperative also made investigative requests. Authorities recommended that use of the water for saunas and washing should be avoided. This was the beginning of a long period during which the environmental impacts were discussed extensively. The public debate focused on the reasons and background causes of environmental problems, especially the water problems. The expertise of the regional environmental authority and their awareness of operating conditions in the mine was called into question. (Tiainen et al., 2014).

Many activities related to Talvivaara's operations were seen as failures. These included the relations with the local population. In 2012, despite the reduction in discharge levels, active public debate continued. The company tried to dispel the rumors by issuing new information. At the end of 2012, the public atmosphere concerning Talvivaara's operations calmed down a bit, but then flared up again when the mine's gypsum sediment pond leaked in November 2012. The leak was located within three days but it could not be immediately stopped. Water with high metal concentrations was discharged from the gypsum sediment pond into the environment for about a week. The regional environmental authority, KaiELY, recommended the surrounding bodies of water should not be used for recreational or household purposes and called for a police investigation of the leak. The public response and criticism were massive and Talvivaara was again one of the main topics in the national media. Environmental NGOs and individual politicians called for the closure of the mine. Criticism included questions about whether the administration possessed sufficient resources and expertise to monitor Talvivaara. In November 2012 Talvivaara received permission to restart metal production which had been stopped due to the leakage. Problems, however, remained unresolved (Tiainen et al., 2014; Sairinen et al., 2017).

In the beginning the central actors were the mine and company management and the local people. The main conflict concerned the breaking of the original promises. The company representatives had promised that the mine would not have any negative impacts on the surrounding environment. This promise was not kept and, as a result, the company lost the confidence of local people (Mononen, 2015). Activism emerged at local, regional as well as national levels. Citizens wrote letters and opinions to the newspapers, and they organized public demonstrations. One of the demonstrations was organized in 2011 by the regional organization (Oulu) of Greens of Finland. In 2012, demonstrations were organized around Finland. Hundreds of complaints, statements, and requests for action were sent to different authorities. The mine's neighbours made 156 environmental complaints to the company in 2011 and the regional authority for environmental issues received almost 300 complaints concerning the application for a uranium extraction permit and updating of the environmental permit (Mononen, 2015).

The exacerbation of environmental problems and announcement of the uranium extraction project led to the creation of a national movement *Stop Talvivaara – for the lakes and rivers*. The movement opposed the Talvivaara mine but also the dramatic growth of the mining industry in general. Stop Talvivaara has been in the vanguard of this opposition. In 2012, the movement demanded through a national petition that the Finnish government cancel and revoke the environmental permit of Talvivaara. In Autumn 2017 this campaign was still ongoing, and more than 65,000 people had signed the petition nationally (Stop Talvivaara, 2017).

As the situation intensified, traditional campaigning organizations like the Finnish Association for Nature Conservation and Greenpeace became involved. The critique of mining gradually expanded to include actions by the authorities concerning mining and political decision-making about mining. The Greens of Finland were part of the government coalition. Government ministries were deeply involved in the matter and they also received strong criticism. Finally, the Talvivaara CEO admitted that mistakes had been made, and said they were sorry. In 2013, there was another leakage of a gypsum pond. Uncertainty over emissions and water system management directly affected the company's economic sustainability, and it went into bankruptcy in 2015. The company is now owned by the state-owned company Terrafame Mining Ltd. It is still uncertain as to how and when this complex, multi-level conflict will be resolved. Economic problems have continued. It is uncertain whether the company will be able to cope with its economic and environmental responsibilities or if it will encounter new problems. Once the confidence with different stakeholders has been lost, it is difficult to get it back.

Talvivaara has had many negative environmental impacts. Short-comings in the management of mining waters and the risk of recurrent wastewater leaks raised questions about the reliability of the mining indus-try and stimulated discussion about the justification of the sector as a whole. Talvivaara is not a unique case. The same questions had been raised earlier about the Outokumpu mine. The problems of water manage-ment were supposed to be taken into account later in other Finnish mines. Environmental technology was promised to be developed and Outokumpu Ltd. and other mining companies had promised that they would concen-trate on protecting the environment and natural resources (Björn, 2016a). But this promise was not kept.

Kylylahti mine – local worries about the lakes

Swedish mining company Boliden Ltd. owns the Kylylahti mine, also located in Eastern Finland (Figure 9.1). Production started in July 2012. The main products are copper, zinc and gold with some cobalt and nickel. Kylylahti is an underground mine, and the area of the mining district is 113 hectares. The mining company owns most of the operation's land; other landowners get small compensation payments. Mining is scheduled for completion in 2021. The centre of Polvijärvi municipality is located 2 kms from the mine and the city of Outokumpu is 20 kms away from the mine. The ore is part of the same Outokumpu deposit. Both the mine and processing plant operate under environmental permits. In this mining project, there was originally a plan to build the processing plant near the mine, but the project never mate-rialized. Instead, in 2010 the mining company bought the old Luikonlahti mine in the municipality of Kaavi which is located 45 kms from the Kylylahti mine (Figure 9.1) and renovated it as a processing plant.

According to a survey connected to the environmental impact assessment in 2006 one of the biggest worries among local people was the potential contamination of Lake Polvijärvi. This was mainly based on the history of the negative environmental impacts of the old Outokumpu mine which was seen as a cautionary example. In addition, the fact, that the mining company was not Finnish, led to some suspicions. However, according to the survey, local people considered the mine a positive initiative because of the potential for new jobs, economic impacts, and new possible inhabitants in the area.

In Kylylahti, the public concern was the impact of mining on ground water. Local people were worried about the ability of Lake Polvijärvi to produce sufficient process water to meet the needs of the mine. Thus, a local environmental movement Pro Polvijärvi was established in 2006 at the time of the environmental impact assessment process. At that time the mining company was planning to build the processing plant in Kylylahti, and to channel its wastewaters to Lake Polvijärvi. Members of the public also expressed concerned about the closure of the mine asking whether the mining company would clean up the site once its life was over or just abandon the mine. The Pro Polvijärvi movement did not oppose mining as such but did want to protect the environment. The water that was to be channelled to lake had to be treated according to regulations, had to be oxygen-rich, and compatible with required discharge levels. According the movement, the oxygen-rich water from the mine to Lake Polvijärvi was in principal a good thing as long as there would not also be harmful emissions.

The mining company's original plan was to take the water from Lake Polvijärvi. Pro Polvijärvi was unsatisfied with this plan. The lake is shallow and the movement speculated that it would not withstand the extraction of water. The mining company and the movement began to collaborate and negotiated for a possible solution. The company came up with a new plan for water usage. It would take water from the pits of the old, closed talc mine nearby. Refined water from the mine would then run to Lake Polvijärvi. According to the representative of the mining company, they have received feedback that the quality of the water in Lake Polvijärvi has improved. In addition, the water circulation was found to be improved. Because the processing plant is located further away, the metal concentrations of the water running to Lake Polvijärvi are low. The discussion about the chemicals has been minor. According to Mononen (2015) the location of the processing plant near the mine would have caused more resistance among the local people in Polvijärvi. But because there were less environmental and water influences than expected, the movement has been quiet.

Although this mining case looks peaceful, some worries have arisen. One may say that the worries were relocated when the processing plant was relocated to Luikonlahti. In summer 2014, the mining company received an environmental permit for the expansion of the enrichment plant. The permit allowed the mill to increase the processing capacity from 550,000 tons of ore to 800,000 tons. In addition, the plant received permission to increase the volume of the enrichment pool, and build a new one nearby. According the company's representative, without the permission the production would soon have to be stopped. The Regional State administrative agency granted the company a permission to start, which meant that the practice might be extended despite the possible complaints.

The modification of the environmental permit led to local opposition. In 2013 local people in Luikonlahti village began to oppose the environmental permit. There was fear of emissions and impacts to the water system: what would happen to lakes nearby? In the water-mixing zone, the nickel content of the water is allowed to exceed the levels of the environmental quality norm. A local resident told in a YLE (Finnish Broadcasting Company) news interview, that 'it seems absurd that the mine could be authorized to run the toxins'. The representative of the mining company, in turn, suggested that the impacts will be reduced significantly by increasing the water recycling facility in the region.

Local residents reported that they were trying to prevent water pollution. People contacted the Stop Talvivaara movement and their intention was to establish a similar movement. Based on the fact that a threat of pollution of the water system was bigger near the Luikonlahti processing plant than near the mine, the Ei Kaavivaaraa movement was established in 2014. This movement used the whole repertoire of protest. For example, it arranged an open meeting for local people in 2015 and an unofficial water-monitoring group was established. This movement also organized a panel discussion to which it invited representatives of political parties, mining professionals and experts on environmental issues. It has also cooperated with village associations and in that way gained more local support. As the movement is worried about not only the local lake, but also a larger water system, it also has gained wide understanding from the owners of summer cottages because some of the lakes are important for recreational use. The movement also reminds everyone about the environmental impacts of the old Outokumpu mine.

Not all mining projects lead to activism or the establishment of a movement. In Eastern Finland, the Pampalo gold mine is a good example. It is a small mine, which started production in 2011. A Swedish company, Endomines AB, owns the mine. Commercial production started in February 2011. The actual area of the mine covers approximately one-third of a 300-hectare mining patent area. Pampalo differs from many other gold mine projects in that cyanide, which is generally seen to be one of the worst environmental risks of mining, is not necessary to enrich the gold (PSV – Maa ja Vesi, 1999). In contrast, mechanical processes and flotation are used. According to Mononen (2012), local people trust the mining company. Despite a clear indication of trust, many of them have raised concerns about potential for harm to a lake nearby and the stream leading to it. If a production accident were to occur, for instance a human error, the discharge would enter the rivers and lake, and spread in many directions and end up far away. The region of Ilomantsi where the Pampalo mine is located is also a special case in regard to the history of conflicts concerning the environment and natural resources since in the past a great deal of criticism has arisen in regard to the utilization of natural resources, environmental conservation and employment. The Hattuvaara movement, for example, which opposed the use of herbicides in intensive forestry in the 1970s and 1980s (Björn 2003) is associated with the history of conflicts concerning environmental and natural resources issue around Ilomantsi (Björn 2006). Nevertheless, to date the mine has not stimulated strong opposition.

Conclusion

This chapter has reviewed three different mining case examples that have given rise to opposition and protest. Our interest has been to find out how and why activist groups have been mobilized in these cases. We asked also which modes of protest and activism can be found and discussed how mining companies and other relevant stakeholders react to these mobilized groups.

The oldest case, Outokumpu, demonstrates that activism focused on the use of natural resources has existed for a long time, and the mining industry is not an outcome of the mining boom of the 2000s when the problems of the Talvivaara mine gained attention. One can say, however, that Talvivaara made mining visible across Finland. The media has an important role in Finnish activism. Not all activism gains wider national attention or turns into a movement. In Outokumpu, the mining company was able to manage the publicity. Activists raised their voices, but the company had its own journal and reporters and it did not listen to local voices. Today information about mining projects and companies spreads widely. The main repertoires of activism and protest are similar: meetings, letters and requests to the authorities and other stakeholders.

In the cases of Outokumpu and Kylylahti rural activism and protest has mainly stayed at the local level. However, now activists have good networks and they have lots of information. Non-local activist groups and movements, like Stop Talvivaara, have shown a wider capacity to act, and wider capacity to enroll actors to their groups through social media. Activism in mining issues has, in general, been framed around economic and environmental impacts. At the local level, this means quality of life and incomes and is based on worries about everyday life. Most of the activism is based on the fear of possible impacts and threats. In Kylylahti, worries were based on the history of negative environmental impacts of Outokumpu mine. The Pro Polvijärvi movement gained attention and the mining company took the worries seriously and negotiated with the movement to find a solution suitable for all actors. In Polvijärvi, near the mine, no new environmental worries have arisen. But near the processing plant, in Luikonlahti, worries about the wastewaters have risen.

The trust between the mining company and the local people is a crucial question. There must be lots of information and cooperation. In Outokumpu and Talvivaara, the attitude of the mining companies was rude in the beginning but changed later. In both cases, dams broke and nearby lakes were contaminated by wastewater. It has been difficult to build up the trust again. In Kylylahti the mining company has listened to local activists and the dialogue with local people has been successful. In the Pampalo case by contrast local activism is largely absent despite some concerns over potential impacts.

The question whether these movements are old or new, is interesting. Stop Talvivaara could easily be considered as old, because it appears as an anti-business movement with one clear aim: closing down the Talvivaara mine and stopping all mining in Finland. However, the motives are not economic; they are environmental and related to quality of life. The movement acts at different scales and helps other movements and activist groups. The Pro Polvijärvi movement was established because of environmental worries as well, but only at the local level and only for one purpose. Also, in Outokumpu, the movement stayed at the local level and did not gain any national publicity because it was a remote place and at that time, the mining company was able to control publicity. Mining activism has characteristics of both old and new movements and will not sit easily in either category. The activism that has emerged can be interpreted as consistent with other rural activism in Finnish rural areas such as that arising from agricultural water pollution. Mining activism in Finland is multifaceted. Rural people do not accept any and all possibilities for local and regional economic benefits, they have and continue to actively defend their environment and the quality of their water in particular.

References

Björn, I. (2000). Kaivos kaupungin yllä. In Björn, I., Immonen, L., & Pennanen, M. (Eds) *Outo kumpulaisuus. Kaivoskaupungin historiaa*. Hämeenlinn, Finland.

Björn, I. (2003). *Ympäristöpolitiikka metsässä?* Joensuu: Joensuun yliopisto, Karjalan tutkimuslaitoksen.

Björn, I. (2006). *Ilomantsin historia*. Keuruu: Otava Publishing.

Björn, I. (2014). Rikkiä, sumppia ja syanidia. Outokummun jätevedet ja lähiluonto. In I. Björn (Ed.) *Ihmeellinen luonto. Kirjoituksia luonnoista. Pohjois-Karjalan historiallisen yhdistyksen vuosikirja 16* (pp. 134–149). Joensuu, Finland.

Björn, I. (2016a). Ympäristötekniikan kehitys Suomen kaivoksilla. Outokumpu antaa esimerkin. *Finnish Quarterly for the History of Technology*, 34, 1, 22–41.

Björn, I. (2016b). Ympäristön puolesta – nuijasodan hengessä. In K. Miettinen & J. Kuhanen (Eds) Ystävällistä viisautta. Professori Harri Siiskosen 60-vuotisjuhlakirja (pp. 17–35). Joensuu, Finland: University Press of Eastern Finland.

Conde, M. & Le Billon, P. (2017). Why do some communities resist mining projects while others do not? *The Extractive Industries and Society*, 4, 3, 681–697.

Huttunen, V. & Sivonen, T. (1974). *Kuusjärven – Outokummun historia*. Joensuu, Finland: Outokumpu Trade and Tourism Committee.

Konttinen, E. & Peltokoski, J. (2004). Ympäristöprotestin neljäs aalto–Eläinoikeusliike ja uuden polven ympäristöradikalismi 1990-luvulla. Jyväskylä: Minerva Kustannus.

Kuisma, M. (1989). *A History of Outokumpu*. Jyväskylä: Gummerus Publishers.

Larsen, S. L. (2008). Place making, grassroots organizing, and rural protest: A case study of Anahim Lake, British Columbia. *Journal of Rural Studies*, 24, 2, 172–181.

Lindholm, A. 2005. *Maailman parantajat. Globalisaatiokriittinen liike Suomessa*. Helsinki: Gaudeamus.

Massey, D. (2008). *Samanaikainen tila*. Tampere: Vastapaino.

McKenzie, F. M. H. (2009). Farms and mines: a conflicting or complimentary dilemma in Western Australia? *Journal for Geography*, 4, 2, 113–128.

Measham, T., Mckenzie, F. H., Moffat, K., & Franks, D. (2013). An Expanded Role for the Mining Sector in Australian Society. *Rural Society*, 22, 2, 184–194.

Moody, R. (2007). *Rocks & Hard Places: The Globalization of Mining*. London: Zed Books.

Mononen, T. (2012). Kaivostoiminnan luonnonvara – ja ympäristökysymykset maaseudulla – esimerkkinä Pampalon kultakaivos. *Maaseudun uusi aika*, 20, 2, 21–36.

Mononen, T. (2015). Jos olisi tavallinen kaivos – Talvivaaran kaivoshankkeen ympäristövaikutukset paikallisten kokemana. *Terra*, 127, 3, 113–124.

Mononen, T. & Suopajärvi, L. (Eds) (2016). *Kaivos suomalaisessa yhteiskunnassa*. Rovaniemi: Lapland University.

PSV – Maa ja Vesi. (1999). *Pampalon kultakaivoksen ympäristövaikutusten arviointiohjelma*. Finland: PSV – Maa ja Vesi.

Rannikko, P. (2003). Oikeudenmukaisuuskysymys suomalaisen ympäristöliikehdinnän aalloissa. In A. Lehtinen & P. Rannikko, (Eds). *Oikeudenmukaisuus ja ympäristö* (pp. 160–180). Helsinki, Finland: Gaudeamus.

Rasimus, A. (2006). *Uudet liikkeet. Radikaali kansalaisaktivismi 1990-luvun Suomessa*. Tampere: Tampere University Press.

Richards, J. (Ed.) (2009). *Mining, Society and a Sustainable World*. New York: Springer. DOI 10.1007/978-3-642-01103-0.

Ryhänen, R. (1960). Outokummun kaivosteollisuuden jätteiden vaikutus vesistöihin. A. Biotyyppi. Pro gradu. Helsinki: Department of Limnology. University of Helsinki.

Saartenoja, A., Törmä, H., Valkosalo, P., & Zawalinska, K. (2007). Talvivaaran kaivoksen aluetaloudelliset vaikutukset Ylä-Savon seutukuntaan, sen kuntiin sekä Rautavaaran kuntaan. Raportteja 21. Helsinki: University of Helsinki, Ruralia-instituutti.

Sairinen, R., Tiainen, H., & Mononen, T. (2017). Talvivaara mine and water pollution. An analysis of a mining conflict in Finland. *The Extractive Industries and Society*, 4, 3, 640–651.

Saleem, H. A. (2009). *Mining, the environment, and indigenous development conflicts.* Tucson, AZ: The University of Arizona Press.

Särkikoski, T. (1999). *A Flash of Knowledge: How an Outokumpu Innovation became a Culture.* Espoo, Finland: Finnish Society for History of Technology.

Scott, J. C. (1985). *Weapons of the Weak: Everyday Forms on the Peasant Resistance.* New Haven, CT and London: Yale University Press.

Singh, I. & Evans, J. (2009). Natural resource-based sustainable development using a cluster approach. In J. Richards, (Ed.) *Mining, Society and a Sustainable World.* (pp. 183–201). Berlin, Heidelberg: Springer. DOI https://doi.org/10.1007/978-3-642-01103-0.

Strijker, D., Voerman, G., & Terluin, I. (2015). *Rural protest groups and populist political parties.* Wageningen, Netherlands: Wageningen Academic Publishers.

Stop Talvivaara, (2017). Website. Retrieved 6 October 2017 from http://stop.viestinta.org/node/393.

Tiainen, H., Sairinen, R., & Mononen, T. (2014). Talvivaaran kaivoshankkeen konfliktoituminen. *Ympäristöpolitiikan ja -oikeuden vuosikirja VII* (pp. 7–76). Joensuu: Institute for Natural Resources, Environment and Society (LYY Institute), University of Eastern Finland.

Urkidi, L. & Walter, M. (2011). Dimensions of environmental justice in anti-gold mining movements in Latin America. *Geoforum, 42,* 6, 683–695.

Vesistön pilaantumista aiheuttavan toimenpiteen rankaisemisesta (1966). Professoreiden Tauno ja Reino Ellilän 14.3.1966 antama lausunto Outokummun kaivoksen jätevesikysymyksistä. Professori Ilmari Melanderin 5.4.1966 antama lausunto samoista kysymyksistä. Joensuu, Finland: Information pamphlet published by Outokumpu Ltd.

Woods, M. (2003). Deconstructing rural protest: the emergence of a new social movement. *Journal of Rural Studies, 19,* 3, 309–325.

Woods, M. (2008). Social movements and rural politics. *Journal of Rural Studies* 24, 129–137.

Woods, M. (2011). *Rural.* London: Routledge. DOI https://doi.org/10.4324/9780203844304.

10 Agritourism in Poland
A new social movement

Grzegorz Foryś

Introduction

New social movements have primarily been discussed with an urban focus. Castells for example concentrates on the impact of capitalist dynamics on the urban space and the role urban social movements play in this process. Castells sees three basic issues in the activity of urban movements: the forms of collective consumption organized by the state, the importance of cultural identity related to territory, and the pursuit of decentralized forms of rule (Castells, 1983, 1997).

But in late modern society, the city is not the only place where new social movements emerge. Mooney (2000) argues that rural social movements are an expression of the fight for preservation of identity by rural residents. This chapter argues that in contemporary rural Poland agritourism has emerged as one such movement. The chapter is based on in-depth interviews involving 20 agritourist farms from Malopolskie Province and 35 leaders of agritourism associations operating in Poland. These interviews were preceded by four focus groups, each with ten people associated with agritourism as owners of agritourism farms, leaders of agritourism associations, experts in agritourism, or state agricultural counselling centres who contribute their expertise to the development of agritourism. The goal of the research was to test the hypothesis that agritourism in rural Poland has the characteristics of a new social movement in terms of goals, human resources, ways of operation, organization, and local orientation. The chapter first discusses rural areas as places for the development of new social movements and discusses the characteristics of the agritourism movement in Poland before asking in what respects the agritourist movement fits the characteristics of a new social movement. The chapter concludes that the hypothesis is confirmed. Agritourism in Poland presents the characteristics of a new social movement.

Rural areas as places for the development of new social movements

There are two lines of analysis which suggest that rural areas are a good place for new social movements to appear. The first is provided by

Mooney (2000, pp. 47–52) who offers several arguments. First, he argues that the development of new communication technologies enables rural residents to overcome their relative isolation, this increases the frequency of their interaction and is conducive to greater collective activity. Second, social categories in rural areas are less varied than those in urban areas and this also promotes the active engagement of rural residents with each other. Third, the resources present in rural areas contribute to social activity. Here, Mooney points to the resources of agriculture in contemporary economy and politics: farmers are a political power able to defend their interests by taking collective action. Fourth, agricultural and rural movements maintain relationships with movements external to rural areas, especially ecological ones, which are attracted to the countryside. Fifth, rural people want access to collective consumption of goods as much as urban people do, so they are inclined to engage in mobilization to create these goods. Sixth, rural people care about protecting their collective identity as much as urban people do. And perhaps even more, since the elements of their rural identity are still present in rural areas, unlike in the town, in rural areas there is a lot to be protected. Last, but not least, rural areas are a better place to resist the colonization of everyday life, as understood by Habermas (2003), than the city. The pressure of the *System* is clearer in rural places and rural social movements manifest active resistance to colonization designs.

Van der Ploeg (2008) provides a second body of analysis identifying factors that indirectly influence the emergence of new social movements in rural areas. He identifies the main tendencies present in contemporary global agriculture and the role played in it by the *Empire* including agribusiness, scientific knowledge, production technologies, engineering, infrastructure, and agricultural institutions. According to van der Ploeg, contemporary agriculture presents two distinct systems. The first system follows the principles of decentralized production and consumption of food, with strong connections between agriculture and local communities. The other system is centralized. It involves specialized agricultural companies that operate worldwide. These constitute the *Empire*. Most important is the industrialization dynamics of global agriculture which generates two effects of key importance: 'repeasantization' and 'deactivation'. Repeasantization arises because industrialization exerts pressure on local and regional agricultural production systems and leads to the marginalization of some agricultural enterprises. The pricing dictates of the *Empire* result in the inability of some to stand up to market competition and, according to van der Ploeg, leads to bankrupt farms. The phenomenon results in the economic degradation of the bankrupt farmers, but on the other hand it allows them to become independent of the market and hence to broaden their autonomy. Industrialization is the macro process generating these effects, but direct causes often appear more mundane. Getting out of the market may well be the result of a reduction in prices, inability to pay off a loan, or failed investments.

Van der Ploeg calls the other process 'deactivation'. Its causes are quite similar to those of repeasantization, but deactivation may also be initiated by institutional decisions such as production quotas which can limit agricultural production in a certain area and thus eliminate some agricultural producers. In such cases agricultural producers may discontinue production activity, sell the farm, and begin again in another sector of the economy. This does not always mean they stop working in rural areas; they can for example function as 'green energy' producers, in rural tourism, or in other non-agricultural activities. The common denominator of both processes is the process by which agricultural industrialization creates tensions between farm owners and the market economy. The primary stakeholders in the fight are so-called 'new' peasants – those that emerge from repeasantization processes. 'New' peasants are individual farmers who under the pressure from the *Empire* have adopted less industrialized and more traditional forms of farming. In new conditions and in a new form they strive to oppose the *Empire*. One might say they are fighting for their independence using local solutions to global problems. Thus, for some of the farmers, it is a fight for their autonomy.

Social movement theory has been applied to explain defiant rural collective behaviour (cf. Edelman, 1999; Foryś, 2008; Gorlach, 2001; Halamska, 1988; Jenkins, 1982; Migdal, 1974, 1975; Paige, 1973; Reed, 2004; Scott, 1976, 1990; Shanin, 1973; Wolf, 1973). But these studies do not refer to the everyday life of rural residents much less to their resistance to *System* pressure as highlighted by Habermas (2003). Instead rural producers participate in the creation of an agriculture model that can compete with the *System*. An organizational and institutional framework is developed with activities that involve a post-productivist or sustainable model of agriculture (Marsden 2003). This kind of activity creates the social basis for new social movements. The owners of small farms operating in accordance with the post-productivist or sustainability model engage in such projects as a modern form of resistance. They strive to retain their identity and place in the world. Rural producers do not try to achieve their goals by using the traditional repertoire of collective action such as protests, rebellions or peasant wars. The fight is not on the streets, in demonstrations or protests, but in everyday activities. Being part of a social movement not only enables the people to defend their collective identity, it also enables them to keep their own farms, maintain the traditional image of the countryside, and uphold post-material values connected with ecology, healthy food, or rural culture. Agritourism is one strategy that gives them a greater chance to achieve those goals.

Rural areas in Poland are a good place for the unfolding of these tendencies. After 1989, when an intensive process of economic and agricultural modernization began, farms were subject to competitive selection. The smallest ones, which were unable to stand up to market competition were pressured to abandon farming or to find some additional sources of income.

After the period of state socialism, at the beginning of this modernization transformation, the mean size of a farm in Poland was 4.5 ha, by 2016 this had increased to around 10.5 ha (GUS 2016).

At the same time the policy of the European Union provided an incentive to keep small farms in production and to diversify their sources of income. The EU provided resources for special programmes to enable income diversification especially ones implementing the idea of sustainable development. Hence, the economic pressure on smaller Polish farms and EU policy combined to create the right conditions for the emergence of new social movements in rural areas and provided a relatively broad social resource base for it.

As part of the Rural Development Programme for 2007–2013, agritourism spending in Poland reached 1.43 billion Polish Zloty (PLN) or roughly €340 million, and in the years 2014–2020 the total funds for rural areas and agriculture are expected to reach EUR 13.6 billion – 8.7 billion from the European Union, and 4.9 billion from the Polish government. This whole amount will go to agriculture and rural areas, and some of it will be allocated to the development of agritourism and other special programmes supporting diversification into non-agricultural rural activities. PLN 100 thousand is available as a subsidy for an individual farm. Micro-enterprise formation and development can receive subsidies between PLN 100 thousand and 300 thousand depending on the number of new jobs created. In addition, the LEADER Plus programme will support the implementation of local action strategies and bonuses paid for establishing non-agricultural economic activities. The owners of agritourism farms can apply for non-returnable subsidies as part of these programmes individually or collectively. There is also the village renewal and development programme, whose beneficiaries are communes or associations who can also support agritourism directly or indirectly.

> Since 2000, there have been unprecedented changes in the demographic structure. For the first time, the influx of people from urban to rural areas is larger than in the other direction. The population of rural areas displays a small increase recently, and this trend is about to continue in the following years.
>
> (Sikorska-Wolak, 2007, p. 14)

To illustrate, in 2015, the rural population was approximately 14.9 million, whereas ten years earlier, it had been about 14.7 million. The proportion of rural residents in the whole community is also growing. In 2000, it was 38.1 per cent; in 2005, it rose to 38.6 per cent, and six years later, to 39.8 per cent (GUS, 2006, 2016). It is estimated that in 2030 the rural population will reach 42.6 per cent (Ministerstwo Rolnictwa i Rozwoju Wsi, 2006). In addition, rural areas cover slightly above

93 per cent of the surface area of Poland, and according to the 2010 census, 54.7 per cent farms in Poland had a farmland size of 5 ha or below (Poczta 2013, p. 14). The rural area in Poland is higher than the EU average and the EU policy concerning them, as well as the high proportion of small farms in Poland are the factors that are conducive to the development of agritourism.

The agritourism sector in Poland

The breakthrough in the development of agritourism in Poland occurred in the early 1990s. Deep changes in this sector were made along with the system transformation at that time. An important institutional change involved the discontinuation of provincial tourist organizations which released agritourism activity from the monopolistic state organizations which had dominated during the socialist period. A second financial change was the influx of national and European resources promoting the development of agritourism. In the beginning, agritourism development was rather scattered but it became more organized in the mid-1990s. Many agritourism organizations were established then, and the circle of people offering agritourism services became more and more integrated. It is worth emphasizing that funds from the European Union granted as part of the PHARE (Poland and Hungary: Assistance for Restructuring their Economies) programme under the Tourin I and Tourin II programmes, which were devoted to supporting tourism contributed greatly to this development.

In 1993, there were approximately 1,000 agritourism farms in Poland and by 2000 this had increased to 5,789. Then between 2002 and 2011 the number of agritourism farms in Poland grew by nearly 20 per cent from 6,546 farms in 2002, to a peak of 10,200 in 2009 (Przezbórska-Skobiej, 2014). There was a considerable reduction in the number of agritourism farms between 2009 and 2010, caused in part by the economic crisis which began in 2008, but their number rose again up to approximately 9,000 in 2016. Most of the participants in the agritourist movement in Poland are owners of relatively small farms. It is noteworthy that the lowest average size of agricultural farms in Poland are in Małopolskie Province (4.04 ha) and Podkarpackie Province (4.77 ha) (GUS, 2016). These provinces also have the largest number of agritourism farms and the most active agritourism associations. In Małopolskie Province, there are about 1,300 agritourism farms and in Podkarpackie Province, approximately 990. Despite this growth the portion of agritourism farms is small. The total number of farms in Poland is about 1,400,000, so agritourism farms represent less than 1 per cent of all farms. The number of guests staying at agritourism farms were: 350 thousand in 2011, 384 thousand in 2014, 452 thousand in 2015 and is estimated to have reached over half a million in 2016 (GUS, 2016).

Agritourism associations are an important organizational aspect of the agritourism sector. It is hard to calculate exactly the number of agritourism associations because of considerable dynamics in this field: some new associations are formed, while some of the existing ones suspend or end their activity. But their number appears to be relatively stable. In 2007, there were around 120 associations in Poland (Zuba, 2007, p. 152). The data of the National Rural Network for 2015 revealed 128.

A significant event in the institutional development of agritourism was the establishment of the Polish Rural Tourism Federation (PFTW), which emerged to represent the whole agritourism sector nationally. There are 45 agritourism associations registered in Poland who are members of the PFTW. Approximately 1,300 agritourism farms are members of these 45 associations. There is a second national association ECEAT Polska. The participants in ECEAT Polska are owners of agritourism farms producing organic food. It is a local branch of an international organization ECEAT (European Centre for Ecological and Agricultural Tourism) founded in the Netherlands. The research showed that the establishment of the Polish branch was motivated by the pursuit on the part of the Dutch to find in Polish rural areas everything that the Dutch rural areas had lost as a result of modernization. It should be added that the second national association, ECEAT Polska, is also a member of PFTW. This is possible because the members of the PFTW are associations, while the members of ECEAT Polska are individual owners of ecological farms who also engage in agritourism. As already mentioned, 35 agritourism associations took part in the study presented here. The study involved 14 associations that were members of PFTW and 19 associations that were not, as well as both nationwide organizations.

The agritourism movement in Poland

Based on the interviews conducted, this section provides a comprehensive characterization of the agritourist movement as viewed by its participants. The purpose is to look at agritourism by referring to the broadest possible set of indicators which could confirm the existence of a movement and allow a characterization of its multiple aspects. This will prepare for the later assessment of the agritourism movement as a new social movement.

The agritourism features identified through the research focus on forms of interaction within the sector; the attitude towards social and cultural change; the organization of the sector; the level of self-awareness; the community of goals and opinions, as well as its spontaneous and distinct activities. This is a similar set of properties as found by Piotr Gliński, when he analysed the ecological movement in Poland (Gliński, 1996).

Internal relations within the sector are relatively frequent and they mostly involve relationships between farm hosts. However, the networks of relations are territorially varied. In areas where there are few agritourism farms or where they are adjacent to the big urban agglomeration of Kraków, interactions among farms are less frequent. Where there are few agritourism farms, it seems to be quite natural that interactions would be low due to the relatively long distances between farms. In addition, being adjacent to a big city makes agritourism farms more likely to focus on providing accommodation for tourists who intend to go sightseeing in Kraków, which they find attractive. The owners of such farms are not as integrated with the values inherent in agritourism as those from places where agritourism is more developed. There are two other tendencies related to interactions within the sector. The first is weak relations between the local agritourism associations. Associations are basically active only in the area where their participants live. The second tendency is the mostly horizontal relations that are found within local associations which results in flat organizational structures.

Agritourism hosts and local associations do interact with a wider institutional environment which includes territorial authorities, agricultural advisory centres, and associations from sectors other than agritourism. These institutions usually have mandates to help develop rural areas, to maintain rural tradition, or to cultivate folk culture. The relations between the local agritourism organizations and these external institutions are largely cyclical. This is connected with the participation of agritourism associations in state-funded (or, more often, EU-funded) rural development programmes. Applying for financial resources and using these resources initiates networks of cooperation between agritourism organizations and the institutional environment. This kind of movement activity shows that it is largely determined by the institutional environment, especially local authorities. On the other hand, it also suggests that the agritourism sector is potentially participating in modernization processes in rural areas. EU funds for the implementation of specific developmental programmes (e.g. the construction of bicycle routes, extension of agritourism farm services, landscape conservation), may directly change the rural environment and undermine its traditional character.

With respect to attitudes to political or cultural change the participants in the agritourist movement have a relatively high level of awareness of the modernization changes taking place in their areas and an understanding of why these changes are necessary. However, despite a relatively high level of acceptance for some modernization processes there is also substantial criticism. Criticism is made mostly of changes that destroy the traditional rural landscape, such as planting vegetation that is foreign to the area around houses, or changes that discourage the existence of small farms or otherwise destroy rural tradition and culture. The agritourist movement serves as a kind of filter for modernization processes, accepting everything that makes it possible to retain rural resources such as nature, organic food, and the rural landscape while criticizing changes that affect the traditional character of the countryside. In general, the agritourist movement can be described as culturally conservative. It attempts to retain everything that determines the essence of the countryside and the rural character, everything that the town has lost, and that is currently at risk of being lost in the countryside as well, due to intensive modernization processes.

The agritourist movement has several unique organizational properties. The basic organizational form is collectivist in nature. This is expressed through properties such as: recruitment based on relationships of friendship, egalitarianism, the lack of rigid rules, the lack of centralized authorities, the lack of hierarchical structure, consensual decision making, rotation in the assignment of responsibilities, limited diversification of social roles, promotion of shared values, and a sense of community. These characteristics are identified as the attributes of an idealized collectivist organization by Rothschild-Whitt (1979) and the organizations of the agritourist movement manifest these features. As a result, the sector is characterized by organizations with a flat organizational structure, the social roles may be exchanged between associations and owners of agritourist farms, and the leadership is task-based.

In addition to private hosts and agritourism associations the research demonstrated the existence of other actors who are important to the agritourism sector but are not represented by formal organizations. They are made up of people who do not work directly in agritourism but closely collaborate with the sector and have the same values connected with the countryside and rural character as agritourist hosts and associations. These are usually the owners of small farms who engage in small-scale agricultural production, mainly organic. Folk artists and folk culture organizers also belong to this group. All these individuals and institutions identify themselves as participants in the agritourist movement.

The unique characteristics of the agritourist organizations are mostly connected with the role played by agritourism associations. The local agritourism associations do not seek to play the role of sector representatives. The principles applying in the associations are shared authority, shared responsibility, equality of the participants, the lack of a clear power structure, and consensual decision-making. But the situation does change as a result of contacts the agritourist movement has with the institutional environment, for example participation in formal relations with local authorities, most often in the context of applying for financial resources for specific projects. In that situation the associations' role and way of operation are different. Formal aspects come to the fore, with the role of leaders (presidents) being of key importance as they are responsible for any entrusted financial resources. As a result, the way of managing the association becomes more formalized. Then we can say that associations do serve as representatives of the movement. Such situations are relatively rare, but they show that agritourism associations may play a twofold role. From day to day they are just another component of the sector, but in extraordinary situations, when they engage in carrying out projects funded with public resources, they assume a representative role. In other words, the role of associations in the agritourist movement is dynamic in character and may even be described as a hybrid.

This is even more evident when we look at the Polish Rural Tourism Federation (*Polska Federacja Turystyki Wiejskiej*, PFTW), which is one of the two nationwide agritourism organizations and the only one that permanently serves as a national representative of the movement. It represents the movement before central state institutions, particularly the Ministry of Agriculture and Rural Development, and the Ministry of Sport and Tourism, which supervises the field of tourism. The PFTW is made up of local associations, not individual agritourism farm owners, and the participation is voluntary. Approximately half of the 130 agritourism associations in Poland are participants of the PFTW. Despite the existence of this one representative nationwide entity the structure of the movement remains flat and the dominant relations are horizontal.

The goals expressed by those active in the sector can be divided into two groups. The primary goals which were reported as most important by the sector's participants were: promoting the farm/village/region, influencing local policy concerning projects to support agritourism, satisfying the expectations and needs of farm tourists, protecting environment and landscape, maintaining the farm as an agricultural operation, sustaining the rural character, stimulating and maintaining interpersonal relations with others, providing educational activities, making agritourism profitable, developing and integrating local communities, and encouraging organic farm production. The less important, secondary goals are improving the standard of agritourism services, improving the knowledge of farm hosts, changing people's attitude to nature, developing mutual cooperation, expanding agritourism, and supporting sustainable development. Both groups of goals include tangible and intangible ones, which makes it difficult to clearly say which goals dominate in the agritourism movement.

There are some owners of agritourism farms who have moved to rural areas in order to pursue post-material goals. For them profits were never the dominant motivation. Hence, newcomers display the strongest attachment to post-material goals and values. They are also the *avant garde* of the agritourism movement. But respondents suggested that the goals of farmers who decide to become hosts has also shifted over time. When beginning to engage in agritourism, the main, or even the only motivation was the expected revenues and the opportunity to sustain the family farm. But after some time, in most cases the revenues prove to be rather low. The hosts can see, however, that agritourism makes it possible to achieve many goals that they had not thought about before and their thinking changed quite dramatically. Post-material goals came to the fore, especially those connected with ecology, maintaining the rural character through rural culture and landscape, interpersonal gratifications, and the development of local community.

The goals and values promoted by agritourism associations are clearly post-material, which is confirmed by the contents of their websites. But the goals declared in the charters of the organizations are more utilitarian and mainly refer to the practical activity connected with organizing and developing agritourism and local communities. Some commodification of values is evident in their activity. The association leaders assert this by stressing that they do pursue post-material values. The presence of goals such as the development of the farm, improving the quality of services, cooperation, or enhancing profits from agritourism are inherent in agritourism as a business and are not viewed by respondents as contradicting post-material values. According to the participants the utilitarian goals are the instruments that allow for the achievement of the post-material goals.

The opinions of movement participants predominantly overlap on issues related to ecology, rural character, and rural culture. Ecology means the need they often mention to care for the natural environment, to keep the landscape unchanged, and to produce organic food. Rural character as perceived as the characteristic rural landscape with fields and typical rural buildings, but most importantly the communal interpersonal relations based on direct contacts, mutual help and support, as well as collaboration within the local community. The last element identified by the movement participants is rural culture. The interviewees mostly meant cultivating local customs and traditions, organizing folk culture (e.g. folk ensembles or rural music bands) and the development of material outputs thanks to the activity of folk artists. Furthermore, the expression of opinions also revealed that the participants treat their agritourist activity as a mission. This was underscored by stating that although the financial motivation for agritourism is common, it is not of primary importance.

Finally, we need to mention the spontaneous and distinct character of the agritourism movement. Spontaneity manifests itself in a rather specific way, in everyday activity of the participants. Their repertoire of activities is not confrontational as is found in many social movements (demonstrations, rallies, riots). They were not particularly intensive in the initial phase of movement development, either. The reason was that there were few conflict situations with the participation of the agritourist movement, and if there was a clash with local authorities, the conflict immediately moved to the negotiation phase and did not trigger such forms of protest. The case of some agritourism farms from the region of Uście Gorlickie (in Małopolskie Province) is a good example of that. Their owners, motivated by the intention to preserve the rural, traditional landscape of the village, lobbied the local authorities to stop the activities aimed at changing its rural character, such as building sidewalks, asphalting driveways, or paving farmyards.

In the initial phase of the movement there was also not much room for spontaneity because it was inspired unintentionally by external factors such as agricultural advisory centres, communal offices, and non-governmental organizations and was immediately placed in a specific legal and formal framework. But spontaneity manifests itself in daily business and social life contacts between the participants, and first of all, in direct democracy by which the movement functions and the solidarity culture is present especially at the association level.

The distinctness of the agritourism movement is visible in several areas. One of the areas is its local character. There are agritourism communities scattered all over the country, which share the same awareness, the same goals, and the sense of being part of an agritourism movement. This leads to another distinct feature: direct relations between movement participants are dominant. Another expression of distinctness is the evolution of the movement, which can be summarized by the phrase 'from individual interests through the community of interests up to social movement'. Finally, the fact that participants of the movement treat it as their lifestyle is unique. This is also visible in their attitude to modernization processes. The movement tries to eliminate the negative effects of modernization, combining the promotion of a certain set of values with practical actions.

Second, activities of the movement are oriented towards the development of social awareness (organizing conferences, printing information and promotion materials, participation in environmental actions, promoting the idea of sustainable development, organizing and participating in conferences, direct contacts with rural tourists, participation in environment protection actions). This is the action repertoire of the agritourist movement. In the case of the agritourist movement there are virtually no protests, happenings or marches, but there are other forms of collective activity, for example organizing picnics, fairs, or festivities. These are opportunities to promote farms and agritourism, but also the values that have been repeatedly mentioned. Collaboration with the owners of Polish ecological farms (the number of which in 2015 was 26.5 thousand according to a report on the state of ecological farming in Poland (IJHARS, 2015) and rural artists are important. They naturally support the agritourist movement, sharing the set of values with its participants. We may even say that they take part in creating the movement.

The agritourism movement in rural areas is important for the rural community. In this context, the activity of the movement in the local public sphere should be highlighted – it takes part in building networks of cooperation and hence contributes to the development of human and social capital. Agritourism activities provide individual and social opportunities for development. Especially important is the role in integrating local communities and enabling rural residents to build a new identity. This aspect is particularly significant in the context of development of late modern society. The general function of the agritourist movement in rural areas is the promotion of sustainable development. This most clearly shows its role in the implementation of state and EU policies concerning Polish rural areas. The practical role of the movement is its contribution to the development of agritourism, thus facilitating the maximization of income for agritourism farms, and when it serves as a guard protecting nature, rural landscape, and rural culture. All the aforementioned functions of the agritourism movement make it an important developmental factor of contemporary rural areas in Poland. It serves an important role, not only maintaining what has been the essence of rural areas so far, but also promoting and implementing values (ecology, healthy lifestyle, organic food, beautiful landscape), which are now the object of interest, not only of rural residents, but also of other people. The role of the agritourist movement is to bring to light these resources in rural areas, which reproduce the rural space, protect the landscape, and give them new significance. Thanks to the acquired financial resources the agritourist movement does not only change the awareness of rural residents, but also the material conditions of their existence.

The agritourism movement as a new social movement

The features of the agritourist movement outlined earlier include a number of properties that justify categorizing it as a new social movement. The goals of a social movement are the element that most clearly determines whether it is viewed as an 'old' or 'new' social movement. According to Offe (1985) and Inglehart (1990), post-material values are the driving force of new social movements. In the case of agritourism the movement is oriented towards culture and the promotion and implementation of post-material values. The significance of post-material goals is growing in the awareness of hosts. Individual hosts have tended to redefine their goals to focus on non-material benefits. The fact that income in agritourism is usually rather low leads to the growing role of other issues in the hosts' awareness: ecology, rural culture, as well as the importance of interpersonal contacts and meeting other people. Collectively, in turn, the dynamics of the values are connected with the development of the whole movement. As the movement became more formalized and organizations were established, the post-material values were included in their statutory goals. They became a permanent component of the agritourist movement, which is an incentive for its new members and supporters.

Based on this collection of goals and opinions, the agritourist movement has developed an ideology which suggests that the movement is relatively mature. Beck (2013) proposed four perspectives for analysing the presence of ideology in a social movement (psychological, interactionist, framing, and structural). Gerlach (2001) identified higher-level ideology (involving views shared by the movement participants) and lower-level ideology (concerning the movement tactic and minor goals).

From Beck's psychological perspective, the role of ideology is to enable actors to attribute meanings and sense to the social world, both individually and collectively. It is used by individuals and social groups as the basis for their own identification. The interactionist perspective treats ideology as an interactive phenomenon, manifested in social movements in periods of activity, and is developed by interactions between the leaders and participants of the social movement and between the movement and its environment. Beck adopts the framing concept suggested by Benford and Snow (2000). An ideology is a frame that articulates, reinforces and transforms the existing beliefs and values into models and objectives of a social movement, at the same time being part of more general culture. In the last concept, Beck presents ideology as a factor permanently reflected in the social order and social structure. This way he shows the process of its institutionalization.

The concepts discussed by Beck can be applied to individuals, to communities, and to social structure. In the case of the agritourist movement, it seems that ideology functions at the individual and the social levels. Owners of agritourism farms seem to be so attached at the individual level to the set of values they promote in the movement that the values become the basis for their self-identification. Responses from farm owners show the relation between individual identity and collective activity within the movement. Moreover, the aforementioned clash of values and goals promoted by agritourism hosts with the expectations and values increasingly popular in the society allows us to treat agritourist ideology as a specific framework rooted in the society, mostly administered by agritourism farms' owners. The content of the ideology is both the care about nature, landscape and tradition, and selected elements of the ethos of the peasantry.

Using the division proposed by Gerlach, we can say that the higher level of ideology is visible in movement participants sharing values such as: protection of the environment and the landscape and sustaining rural culture and tradition, while the lower level relates to: certain components of the peasant ethos and specific goals connected with the promotion of the region, tourism, obtaining profits from running the farm, or the directions of development of farms.

In addition, the action repertoire of the agritourism movement is oriented towards the development of social awareness. There are virtually no protests, happenings or marches, but there are multiple forms of collective activity. These are opportunities to promote farms and agritourism, and also the movement's post-material values. The relations with the institutional environment also have a political character. This is manifested in situations of direct collaboration with local and regional authorities when carrying out developmental projects in rural areas, but also in conflict situations. In the case of conflicts, the movement does not adopt protest activities, it proceeds to the negotiation phase instead.

The organizational structure of the agritourism movement is not hierarchical. It mostly involves direct horizontal relations and the principles of direct democracy and solidarity. It seems that the form of the organizations results from the movement's local orientation, which is the effect of the tactic of acquiring resources, mostly financial ones, locally. This leads to intensive but cyclical collaboration with the closest environment and to vague boundaries of the movement against the background of other entities.

Finally, the movement is locally oriented, and its human resources originate from the middle class and groups that are outside the system (the unemployed, pensioners, or homemakers).

All these features lead to the conclusion that the movement can legitimately be called a new social movement. This does not mean that this statement is beyond any doubt. Several characteristics seem to point in the opposite direction. The permanent profit orientation is a material not a post-material value. However, it is not dominant and in time is replaced by the sense of mission concerning the promotion of post-material values connected with rural identity. Second, the relationships and the sense of being part of the movement are weak where agritourism is not well developed. Yet, this seems natural and only shows the variety of activity of different movement participants. And finally, the agritourist movement is strongly determined by the institutional environment (local authorities, agricultural advisory centres, European funds, or state and EU policies towards agriculture). This evokes the question of whether it is an independent entity. But these counter tendencies seem to be a specific characteristic of the movement rather than features that suggest it is an old social movement or simply a standard business sector.

Conclusion

Rural areas in Poland have many resources upon which a rural social movement has been built including unchanged landscape, organic produce, a peasant ethos, and the remains of communal social bonds which are in decline in countries where agriculture has been completely technicized and dehumanized. All these resources gained special importance in the context of the Polish society's transition from the modern to the late modern phase, in which post-material values are especially significant. They are shared by many participants of society and largely occur in rural areas. Empirical study involving the owners of agritourism farms and the leaders of associations confirmed the hypothesis that the agritourist movement in Poland is a new social movement whose activity relates to the rural land base. Transformation processes in Poland, involving these areas and agriculture, combined with EU and state policies, have created a structural and institutional context that promotes this rural movement. This context is composed of the relatively large group of small farm owners who are weakly attached to the agricultural market, EU policy aiming at the diversification of agriculture and the income sources of rural residents which are encouraging the establishment of agritourism farms, and the fact that numerous natural resources are still present in Polish rural areas. However, the image of the countryside presented here may be undermined by increased commodification of rural resources through the development of rural consumption activities. This is one of the main developmental dilemmas that Polish rural areas will face in the future.

Acknowledgement

This research project was financed by National Science Centre (Poland). Program OPUS 8, No: 2014/15/B/HS6/01228.

References

Beck, C. J. (2013). Ideology. In D. Snow, D. della Porta, B. Klandermans, & D. McAdam (Eds) *The Wiley-Blackwell Encyclopaedia of Social and Political Movements* (pp. 586–590). Oxford: Wiley-Blackwell.

Benford, R. D. & Snow, D. A. (2000). Framing processes and social movements: An overview and assessment. *Annual Review of Sociology*, 26, 1, 611–639.

Castells, M. (1997). *Power of Identity*. London: Wiley-Blackwell.

Castells, M. (1983). *The City and the Grass-Roots. A Cross-cultural Theory of Urban Social Movements*. Berkeley, CA: University of California Press.

Edelman, M. (1999). *Peasants Against Globalization: Rural Social Movement in Costa Rica*. Stanford, CA: Stanford University Press.

Foryś, G. (2008). *Dynamika sporu. Protesty rolników w III Rzeczpospolitej*. Warszawa: Wydawnictwo Naukowe Scholar.

Gerlach, L. P. (2001). The structure of social movements: Environmental activism and its opponents. In J. Arquilla & D. Ronfeldt (Eds) *Networks and Netwars: The Future of Terror, Crime, and Militancy* (pp. 289–309). Santa Monica: Rand.

Gliński, P. (1996). *Polscy Zieloni. Ruch społeczny w okresie przemian*. Warszawa: Instytut Filozofii i Socjologii PAN.

Gorlach, K. (2004). *Socjologia obszarów wiejskich. Problemy i perspektywy*. Warszawa: Wydawnictwo Naukowe Scholar.

GUS – *Główny Urząd Statystyczny*. (2006). *Rocznik Statystyczny 2006*. Warszawa: Główny Urząd Statystyczny.

GUS – *Główny Urząd Statystyczny*. (2016). *Rocznik Statystyczny 2016*. Warszawa: Główny Urząd Statystyczny.

Habermas, J. (2003). *The Theory of Communicative Action, Volume 2: Lifeworld and System: A Critique of Functionalist Reason*. Boston, MA: Beacon Press.

Halamska, M. (1988). Peasant Movements in Poland, 1980–1981: State Socialist Economy and the Mobilization of Individual Farmers. In B. Misztal & L. Kriesberg (Eds). *Research in Social Movements, Conflicts and Change, V.10* (pp. 147–160). London: JAI Press.

IJHARS – Inspekcja Jakości Handlowej Artykułów Rolno-Spożywczych. (2015). *Raport o stanie rolnictwa ekologicznego w Polsce*. Warszawa: IJHARS.

Inglehart, R. (1990). *Culture Shift in Advanced Industrial Society*. Princeton, NJ, Princeton University Press.

Jenkins, J. C. (1982). Why do peasants rebel? Structural and historical theories of modern peasant rebellions. *American Journal of Sociology*, 88, 3, 487–514.

Migdal, J. S. (1974). *Peasants, Politics and Revolution. Toward Political and Social Change in the Third World.* Princeton, NJ: Princeton University Press.

Ministerstwo Rolnictwa i Rozwoju Wsi. (2006). *Programu Rozwoju Obszarów Wiejskich na lata 2007–2013.* Warszawa: Agrotec Spa, Ministerstwo Rolnictwa i Rozwoju Wsi.

Mooney, P. H. (2000). Specifying the 'rural' in social movement theory. *Polish Sociological Review*, 129, 35–55.

Mooney, P. H. (2004). Democratizing rural economy: Institutional friction, sustainable struggle and the cooperative movement. *Rural Sociology*, 69, 1,76–98.

Offe, C. (1985). New social movements: Challenging the boundaries of institutional politics. *Social Research*, 52, 4, 817–868.

Paige, J. (1975). *Agrarian Revolution.* New York: Free Press.

Poczta, W. (Ed.). (2013). *Gospodarstwa rolne w Polsce na tle gospodarstw Unii Europejskiej – wpływ WPR.* Warszawa: GUS, Zakład Wydawnictw Statystycznych.

Przezbórska-Skobiej, L. (2014). *Agroturystyka w Polsce na tle pozostałych krajów Unii Europejskiej.* Warszawa: Materiały Konferencyjne: Rolnictwo, gospodarka żywnościowa, obszary wiejskie – 10 lat w UE. Retrieved 17 October 2014 from http://10eu.wne.sggw.pl/wp-content/uploads/Przezborska-Skobiej_Lucyna.pdf.

Reed, M. (2004). The mobilization of rural identities and the failure of the rural protest movement in the UK, 1996–2001. *Space and Polity*, 8, 1, 25–42.

Rothschild-Whitt, J. (1979). The collectivist organization: An alternative to rational-bureaucratic models. *American Sociological Review*, 44, 4 (Aug), 509–527.

Scott, J. C. (1976). *The Moral Economy of Peasant.* New Haven, CT: Yale University Press.

Scott, J. C. (1990). *Domination and the Arts of Resistance: Hidden Transcripts.* New Haven, CT and London: Yale University Press.

Shanin, T. (1973). Peasantry as a political factor. In T. Shanin (Ed.). *Peasants and Peasant Societies* (pp. 238–263). Middlesex: Penguin Books.

Sikorska-Wolak, I. (2007). Społeczno-ekonomiczne przesłanki kształtowania funkcji turystycznych obszarów wiejskich. In I. Sikorska-Wolak (Ed.), *Turystyka w rozwoju obszarów wiejskich* (pp. 13–28). Warszawa: Wydawnictwo SGGW.

Wolf, Eric R. (1973). *Peasant Wars of the Twentieth Century.* New York: Harper and Row.

Zuba, J. & Zuba, M. (2007). Analiza stopnia integracji wybranych stowarzyszeń agroturystycznych w krajowej federacji agroturystycznej. *Annales Universitatis Mariae Curie-Skłodowska*, 62, 1, 151–161.

Index

Page numbers in **bold** denote tables, those in *italics* denote figures.